Lecture Notes in Electrical Engineering

Volume 77

For further volumes:
http://www.springer.com/series/7818

Ali Rostami · Hassan Rasooli · Hamed Baghban

Terahertz Technology

Fundamentals and Applications

Prof. Ali Rostami
Nanophotonics
School of Engineering-Emerging
 Technology
University of Tabriz
Bolv. 29, Bahman
51666 Tabriz
Iran
e-mail: rostami@tabrizu.ac.ir

Hassan Rasooli
School of Engineering-Emerging
 Technology
University of Tabriz
Emam Khomeini Bolv. 29, Bahman
51666 Tabriz
Iran
e-mail: hrasooli@tabrizu.ac.ir

Hamed Baghban
Faculty of Electrical and
 Computer Engineering
University of Tabriz
Emam Khomeini Bolv. 29, Bahman
51666 Tabriz
Iran
e-mail: h-baghban@tabrizu.ac.ir

ISSN 1876-1100
e-ISSN 1876-1119

ISBN 978-3-642-15792-9
e-ISBN 978-3-642-15793-6

DOI 10.1007/978-3-642-15793-6

Springer Heidelberg Dordrecht London New York

© Springer-Verlag Berlin Heidelberg 2011

This work is subject to copyright. All rights are reserved, whether the whole or part of the material is concerned, specifically the rights of translation, reprinting, reuse of illustrations, recitation, broadcasting, reproduction on microfilm or in any other way, and storage in data banks. Duplication of this publication or parts thereof is permitted only under the provisions of the German Copyright law of September 9, 1965, in its current version, and permission for use must always be obtained from Springer. Violations are liable to prosecution under the German Copyright law.

The use of general descriptive names, registered names, trademarks, etc. in this publication does not imply, even in the absence of a specific statement, that such names are exempt from the relevant protective laws and regulations and therefore free for general use.

Cover design: eStudio Calamar, Berlin/Figueres

Printed on acid-free paper

Springer is part of Springer Science+Business Media (www.springer.com)

Preface

The lack of reliable, high-power, room-temperature terahertz sources and efficient detectors in the past several decades has caused the researchers do not extensively explore the material interactions related to the terahertz spectral region that was once dubbed. However the recent progress in developing of new terahertz continuous-wave and pulsed sources along with user-friendly detectors advent the terahertz-wavelength applications in many fields. For realization of the full potential of THz applications, wide-bandwidth, highly efficient sources and detectors must be developed. Historically, the fields of astronomy and chemistry have been a driving force for developing sources and detectors in the terahertz regime. Even though the applications of terahertz have yet to be used widely, much research is being directed toward the development of terahertz sources and detectors, particularly for applications in medical imaging and security scanning systems.

The electromagnetic waves at frequencies in the region of the electromagnetic spectrum between 300 GHz 3×10^{11} Hz) and 3 THz (3×10^{12} Hz), corresponding to the submillimeter wavelength range between 1 mm (high-frequency edge of the microwave band) and 100 mm (long-wavelength edge of far-infrared light) is called terahertz radiation. This radiation can pass through clothing, paper, wood, masonry, plastic, ceramics and penetrates fog and clouds, but this not apply to metal or water. Terahertz radiation is emitted as a part of the black body radiation from anything with temperatures greater than about 10 K where observations at these frequencies are important for characterizing the 10–20 K cold dust in the interstellar medium.

The generation and detection of radiation in the terahertz range is resistant to the commonly employed techniques in the neighboring microwave and infrared frequency bands where the use of solid state devices has been hampered for the reasons of transit time of charge carriers being larger than the time of one oscillation period of terahertz radiation, and the energy of radiation quanta being substantially smaller than the thermal energy at room temperature and even liquid nitrogen temperature.

Generally there are two types of methods to generate and detect THz signal. There are optical method such as Austin switch, Photomixing, Optically pumped THz lasers and electronic method including Quantum Cascade Lasers, Gunn diodes, Schottky diode frequency multipliers and so on. The THz radiation and detection based on the two-dimensional electron gases (2DEG) oscillation of the field-effect transistor are drawing much attention for its ability to provide frequency–voltage tunable, compact and room temperature workable THz source and detectors. The reported results show that the gated FET with high mobility and smaller channel length results in good performance of the THz radiation source and detector. For example silicon nanowire MOS transistor (SNFET) shows great potential in the future for THz integrated circuit applications.

Today for coherent generation and detection of wideband THz radiation, photoconducting dipole antennas and electro-optic (EO) materials are used in particular for THz spectroscopy and imaging systems. While conventional photoconducting dipole antennas have superior sensitivity, there exists bandwidth limitation and a speed versus sensitivity trade-off. Although the 60 THz detection at low-temperature GaAs photoconducting dipole antennas has been reported, non existing high signal-to-noise ratios and having high-frequency roll-off may limit their use in mid-IR spectroscopic applications. In addition, the low-temperature signal magnitude for GaAs was less than that of reported using ZnTe. EO crystals of ZnTe have shown high-frequency performance up to 30 THz, and the organic crystal DAST has shown performance up to 20 THz. However, both of these EO crystals exhibit wide gaps in their frequency responses that are due to absorption from lattice vibrations and to a phase mismatch between the THz beam and the probe beam in the EO sensor.

Quantum structures, fabricated by molecular beam epitaxy (MBE) can be used for the generation and detection of THz radiation. The THz QCL (quantum cascade laser) as a compact generator also THz quantum well infrared detectors (QWIPs) or heterojunction interfacial work function internal photoemission detectors (HEIWIPs) as detectors are developed which can be engineered by varying the widths of the wells/barriers and the doping profile. Alternatively, one can rely on the properties of carriers along the structure layers for detection of THz radiation using the excitation of plasma waves or non-uniform electron heating in a two-dimensional electron gas.

Some of the viable sources of terahertz radiation are: the gyrotron, the backward wave oscillator, far infrared laser, quantum cascade laser, free electron laser, synchrotron light sources, photomixing sources, single-cycle sources and optical rectification. Among species of terahertz sources, the optically pumped terahertz laser (OPTL) system may be preferable because of its operational simplicity, high signal-to-noise ratio, and ability to use conventional, room temperature detectors. OPTL is in use around the world, primarily for astronomy, environmental monitoring, and plasma diagnostics. Short-pulse terahertz systems are used in time-domain spectroscopy to understand biological processes and to create two or three-dimensional images. The choice of a terahertz source will determine the type

Table 1 Techniques for generating terahertz radiation

	Optically pumped terahertz lasers	Time domain spectroscopy	Backward wave oscillators	Direct multiplied sources	Frequency mixing
Average power	>100 mW	~1 µW	10 mW	mW–µW (decreasing w/increasing frequency)	Tens of nanowatts
Usable range	0.3–10 THz	−0.1–2 THz	0.1–1.5 THz	0.1–1 THz	0.3–10 THz
Tunability	Discrete lines	N/A	200 GHz	~10–15% of center frequency	Continues
Continues wave/pulsed	CW or pulsed	Pulsed	CW	CW	CW
Turnkey systems available	Yes	Yes	No	Yes	Yes

of detection scheme required. Table 1 shows some of the techniques for generating terahertz radiation.

The next generation of detection systems will be multi-modal which will be used for both imaging and detection. The absorption characteristics of terahertz radiation vary greatly from material to material, and this property can be used to create images. Different materials absorb different frequencies by molecules vibrating against each other. Because of some materials being hard to identify this can be used for detection. For example a white powder may be a narcotic or remnants of explosive material, or it may simply be talcum powder or sugar. One of the ways to identify this type of materials is to consider using the absorption property of electro-magnetic energy. For instance, the measured spectrum of Semtex-H plastic explosive at a distance of a metre allows for rapid and reliable identification.

Although terahertz radiation has the potential to revolutionize certain aspects of medical imaging, the problem is the lack of practical detection technology. It is worth to mention that the scientists have developed a detector based on a carbon nanotube transistor that can sense small numbers of terahertz photons. A component known as a two-dimensional electron gas (2DEG) absorbs THz radiation. A single electron carbon nanotube transistor which acts as a switch is laid on top of the 2DEG. By absorption of the terahertz tradition through the 2DEG the transistor voltage will be switched. Experimental and theoretical results clearly indicate that nanometer transistors are promising candidates for a new class of efficient THz detectors. Also the properties of magneto-plasmons in the two-dimensional electron system can be applied for detection of terahertz radiations. For instance, InGaAs/InAlAs Field Effect Transistors in quantizing magnetic field was used for the terahertz radiation detection. The detection is accomplished based on the rectification of the terahertz radiation by plasma waves.

Heterodyne detectors are the most common terahertz detectors although nowadays focuses have been made toward implementation of direct detectors in particular for applications that do not require ultra-high spectral resolution. Some of the most important direct THz detectors are: GaAs–Schottky diodes, conventional bolometers, composite bolometers, microbolometers, hot electron bolometer detector, Golay cells, an acoustic bolometer and finally a fast calorimeter.

A broadband terahertz detector can be realized based on the following idea: if terahertz radiations can be generated from light then it can also be converted back to light. Remember that in the world of optics, there is a range of high-speed, sensitive detectors that can operate at room temperature. For an example a detector was introduced that can detect terahertz radiations indirectly by detecting the light generated when the DAST crystal is exposed to terahertz radiation.

For terahertz time domain spectroscopy (THz-TDS), conventional detectors such as bolometers are not suitable because they can only measure the total energy of a terahertz pulse, rather than its electrical field over time. Instead the following two methods are applied: photoconductive sampling and electro-optical sampling. In both of these methods an ultrashort laser pulse is fed to the detector along with simultaneously applying a terahertz pulse. As a result of this, depending upon whether the detection pulse with the electric field of the THz pulse being high or low, the detector will produce a different electrical signal.

Ultimately, the terahertz detection is possible using both passive and active methods. The passive detectors are effective due to the temperature difference between human body and environment. While the active detectors are cheap but require high power sources. For this reason they are yet to be used with full potential.

Among the most important applications of the terahertz technological under development we mention the followings: medical imaging (terahertz radiation is able to penetrate deep into many organic materials without causing any damage associated with ionizing radiation such as X-rays and can also be used as a cancer detector), security (terahertz radiation can be used in surveillance such as security screening, uncovering concealed weapons, non-destructive detection of narcotics or stimulants in mail, remotely), communications (space and satellite communications), and industrial applications (quality, sensing, monitoring, and process control). Finally as far as the scientific research is concerned we can briefly name: chemistry and biochemistry measurements, study of the complex dynamics involving condensed-matter in high magnetic fields physics, molecular recognition and protein folding, submillimetre astronomy, and detection of murals hidden beneath coats of plaster or paint.

In this book we have only briefly concentrated on the ever expanding world of terahertz where the scope of potential and actual applications cannot be underestimated. Our primary interest lies on the detection and generation of terahertz radiation using optoelectronic quantum devices. In the first chapter of this book, we review the terahertz technology and its associated scientific achievements as important as is in today's world. In the second chapter, the terahertz and infrared quantum photodetectors is studied where its importance is emphasized. Finally, in the third chapter the terahertz and infrared sources based on quantum cascade lasers along with fresh ideas will be analyzed and discussed.

Ali Rostami
Hassan Rasooli
Hamed Baghban

Contents

1. **An Overview of the Technological and Scientific Achievements of the Terahertz** 1
   - 1.1 Introduction 1
   - 1.2 Terahertz Terminology 1
   - 1.3. Terahertz Applications and Opportunities 2
     - 1.3.1 Space 2
     - 1.3.2 Plasma Fusion Diagnostics 4
     - 1.3.3 Spectroscopy 4
     - 1.3.4 Industrial 6
     - 1.3.5 Communication 7
     - 1.3.6 Simulating the Radar Scattering Signatures 7
     - 1.3.7 T-Ray Imaging 8
     - 1.3.8 Medical 11
     - 1.3.9 Physics 12
     - 1.3.10 Chemistry and Biology 14
   - 1.4 Terahertz Components 16
     - 1.4.1 THz Detectors 16
     - 1.4.2 THz Sources 40
   - 1.5 Theoretically Investigation of Terahertz-Pulse Generation and Detection 62
     - 1.5.1 THz Pulse Generation and Detection Using Photoconductive Antenna (PC) 63
     - 1.5.2 THz Pulse Generation and Detection Using Other Methods 65
     - 1.5.3. THz Radiation from Bulk Semiconductor Microcavities 69
   - 1.6 THz Detectors and Sources with Organic Materials 70
     - 1.6.1 Conjugated Semiconducting Polymers 70
     - 1.6.2 Organic EO Polymers 71
     - 1.6.3 Organic EO Crystals 72

		1.6.4	Solutions Containing Polar Molecules	78
	References			78

2 Terahertz and Infrared Quantum Photodetectors ... 91
- 2.1 Introduction ... 91
- 2.2 Detector Principles ... 92
  - 2.2.1 Noise Affects ... 93
  - 2.2.2 Background Limited IR Performance ... 95
  - 2.2.3 Quantum Cascade Detectors ... 95
  - 2.2.4 Effects of Number of Periods and Doping Density on the Detector Parameters ... 101
  - 2.2.5. Quantum Dot Infrared Photodetectors ... 103
- 2.3 Terahertz and Infrared Quantum Cascade Detectors ... 104
  - 2.3.1 Dual Color Mid-Infrared Quantum Cascade Photodetector in a Coupled Quantum Well Structure ... 107
  - 2.3.2 A Dual-Color IR Quantum Cascade Photodetector with Two Output Electrical Signal ... 110
- 2.4 Terahertz Quantum Well Photodetector Based on Two-Photon Absorption ... 115
- 2.5 Quantum Dots THZ-IR Photodetector ... 120
  - 2.5.1 An Overview of Quantum Dot ... 120
  - 2.5.2 An Overview of Quantum Dots Photodetectors ... 123
  - 2.5.3 Ultra-High Detectivity Room Temperature THZ-IR Photodetector Based on Resonant Tunneling Spherical Centered Defect Quantum Dot ... 134
  - 2.5.4 Terahertz Photodetector Based on Intersublevel Optical Absorption in Coupled Quantum Dots ... 149
- 2.6 Terahertz and Infrared Photodetector Based on Electromagnetically Induced Transparency ... 152
  - 2.6.1 Electromagnetically Induced Transparency Phenomena ... 152
  - 2.6.2 EIT-Based Photodetection ... 159
  - 2.6.3 Terahertz Quantum Cascade Photodetector Based on Electromagnetically Induced Transparency ... 178
- References ... 183

3 Terahertz and Infrared Quantum Cascade Lasers ... 191
- 3.1 Introduction ... 191
- 3.2 Quantum Cascade Laser Principles ... 191
  - 3.2.1 Radiative and Non-radiative Transitions in Semiconductor Heterostructures ... 192
  - 3.2.2 Resonant Tunneling Transport ... 196
  - 3.2.3 Quantum Cascade Lasers ... 197
  - 3.2.4 Optical Gain ... 198

		3.2.5	Threshold Current	199
		3.2.6	Losses	199
		3.2.7	Slope Efficiency	200
	3.3	Terahertz Quantum Cascade Lasers		200
		3.3.1	Terahertz QCL Structures	201
		3.3.2	Drude Model	207
		3.3.3	Terahertz-Waveguide	208
		3.3.4	Distributed Feedback QCLs	210
	3.4	Analysis of Transport Properties of THz QCLs		212
	3.5	High Power QCLs		215
		3.5.1	Photonic Crystal DFB QCLs	217
	3.6	Dual-Wavelength Generation Based on Monolithic THz-IR QCL		219
		3.6.1	Linear Frequency-Doubling in IR QCL	219
		3.6.2	Dual Color Terahertz QCL	226
	References			234
Index				239

Chapter 1
An Overview of the Technological and Scientific Achievements of the Terahertz

1.1 Introduction

Due to the importance of terahertz radiation in the past several years in spectroscopy, astrophysics, and imaging techniques namely for biomedical applications (its low interference and non-ionizing characteristics, has been made to be a good candidate to be used as a powerful technique for safe, in vivo medical imaging), we decided to review of the terahertz technology and its associated science achievements. The review consists of terahertz terminology, different applications, and main components which are used for detection and generation of terahertz radiation. Also a brief theoretical study of generation and detection of terahertz pulses will be considered. Finally, the chapter will be ended by providing the usage of organic materials for generation and detection of terahertz radiation.

1.2 Terahertz Terminology

The term 'terahertz' has been applied in different aspects: the frequencies below the far infrared, the frequency coverage of point contact diode detectors [1], the spectral line frequency coverage of a Michelson interferometer [2], the resonant frequency of a water laser, and now terahertz is applied to submillimeter-wavelength range between 1000 and 100 µm (300 GHz–3 THz) [3]. According to the methodology (bulk or modal-photon or wave) we may use the terms far-IR or sub-millimeter. The THz range of the electromagnetic spectrum was called the THz gap because until recently there were no suitable THz signals sources and detectors so the THz portion of the spectrum—sandwiched between traditional microwave and optical technologies—approximately has not been practically used (Fig. 1.1). The terahertz technology was limited to the high-resolution spectroscopy and remote sensing areas. Recently, developed researches have been done trying to fill this gap. Progressing in industrial production of terahertz sources and detectors introduces new instrumentation and measurement systems. THz radiation is having

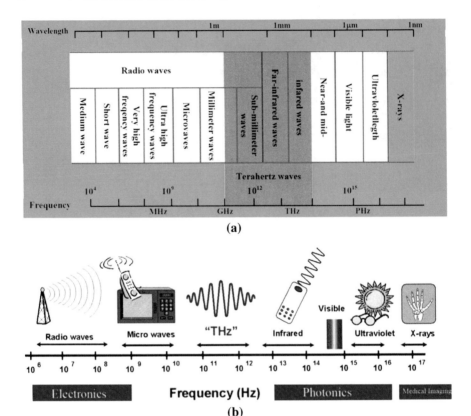

Fig. 1.1 a Frequency and wavelength spectra of electromagnetic waves and **b** applications [4]

widespread applications in medicine—the medical community is getting interested in T-ray imaging, microelectronics, agriculture, forensic science, and many other fields. In fact, the universe is bathed in terahertz energy originated from big bang and most of it going unnoticed and undetected.

1.3 Terahertz Applications and Opportunities

Terahertz radiations have shown unique properties and these radiations provide a variety of applications and opportunities in different fields. Here, we introduce some of these aspects briefly.

1.3.1 Space

Interstellar dust cloud spectral that range from 1 mm to 100 μm (14–140 K below the ambient background on Earth), has made astronomers focused on terahertz

1.3 Terahertz Applications and Opportunities

detection technology (Fig. 1.2). In Fig. 1.2 the radiation spectral of interstellar (dust, light and heavy molecules), a 30-K blackbody, and the 2.7-K cosmic background are shown [3, 5]. Although interstellar dust clouds emit many individual spectral lines, only a few of these lines have been identified. Mapping the terahertz bands with high resolution is needed to eliminate the affect of spectral line clutter and atmospheric absorption. Observing the spectral energy distributions in galaxies show that approximately one-half of the total luminosity and 98% of the photons emitted since the Big Bang fall into the submillimeter and far-IR [6]. Submillimeter detectors allow us true probe into the early universe, star forming regions and many other abundant molecules, e.g., water, oxygen, carbon monoxide, and nitrogen. This type of probing requires two characteristics: high resolving power (large apertures) and high spectral resolution (1–100 MHz). For lower and upper terahertz bands, heterodyne detectors and direct detectors are respectively required.

Many of the interstellarly abundant radiation spectra are found in planetary atmospheres with background temperature varying from tens to hundreds of Kelvin. Monitoring of thermal emission lines of stratospheric and upper tropospheric gases allows studying important atmospheric phenomena such as ozone destruction, global warming, etc. High resolution heterodyne receivers have revealed spectroscopic signatures of these gases at submillimeter wavelengths between 300 and 2500 GHz [7, 8]. Since water and oxygen are of the strongest absorbers for terahertz radiations in the lower stratosphere, longer millimeter waves are required for atmospheric chemical probing.

Observation of asteroids, moons and comets is another application for terahertz detectors which gives a deep insight about the evolution of our solar system [3, 9].

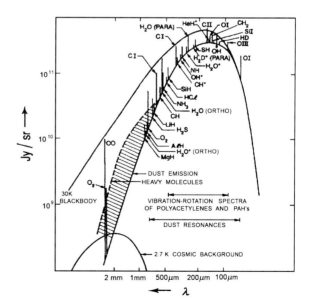

Fig. 1.2 Radiation spectrum of 30-K blackbody, typical interstellar dust and key molecular line emissions in the submillimeter [3, 5]

It is not unreasonable to suppose that the first detection of planets containing atmospheric conditions (temperature, pressure, composition) suitable for extraterrestrial life forms will be confirmed by submillimeter-wave remote detection and terahertz spectroscopy [10].

1.3.2 Plasma Fusion Diagnostics

Terahertz technology can be applied in the plasma fusion diagnostics. Using a narrow-band radiometer pointing along a radial line of the plasma core, one can infer the equivalent blackbody intensity and therefore the temperature of the plasma [11]. Another plasma fusion characteristic is electron temperature fluctuations in the plasma core which appears as white noise riding on top of the thermal electron noise. This additional output noise can be separated out using interferometric techniques which utilizes radiometers that involves heterodyne imaging systems at millimeter and submillimeter-wave frequencies [12].

1.3.3 Spectroscopy

Spectroscopy is one of the interesting applications of terahertz technology which rises from the strengths of the emission or absorption specifications for the rotational and vibrational excitations of the molecules.

Rapid scan and gas identification systems such as FAst Scan Submillimeter Spectroscopic Technique (FASSST) and optical pulse terahertz time-domain spectroscopy devices are of state of the art applications for terahertz spectroscopy [13, 14]. Figure 1.3 illustrates the block diagram of the FASSST system which uses a voltage tunable backward wave oscillator (BWO) as a primary source of radiation, but uses fast scan ($\sim 10^5$ Doppler limited resolution elements/s) and optical calibration methods rather than the more traditional phase or frequency lock techniques.

In order to develop different methods to directly monitor complex molecule dynamics in real time it has been suggested that an atomic level picture of the concerted motions of polypeptide chains and DNAs may be accessible through accurate measurement of low-frequency vibrational spectra. These vibrations are expected to occur in the terahertz frequency regime. Detection of DNA signature through dielectric resonances (phonon absorption) is another interesting consequence which have became realizable due to conceivable measurement and rapidly identification of diverse spectral signatures by terahertz spectroscopy instruments [16, 17]. DNA transmissions are well-defined resonances over a wide range of frequency spectrum (from a few GHz to several THz) as depicted in Fig. 1.4 for Herring and Salmon DNA transmissions.

1.3 Terahertz Applications and Opportunities

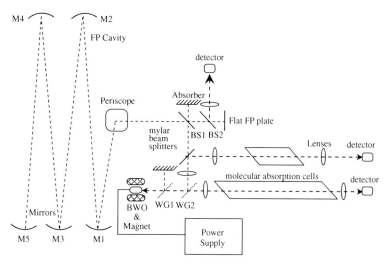

Fig. 1.3 Block diagram of the FASSST system [15]

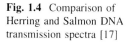
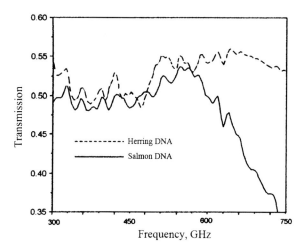

Fig. 1.4 Comparison of Herring and Salmon DNA transmission spectra [17]

Also, there are unique properties of terahertz radiation that make it an interesting technique in the pharmaceutical and security applications. Many materials are semitransparent to terahertz radiation. This property of materials allows the terahertz radiation to penetrate through many everyday physical barriers such as typical clothing and packing materials with modest attenuation. An example of terahertz spectroscopy for common clothing materials is presented in Fig. 1.5. This is because of low photon energies and non-ionizing nature of terahertz radiation which allows THz waves to be used in non-invasive

Fig. 1.5 Terahertz absorption spectra of common clothing materials [18]

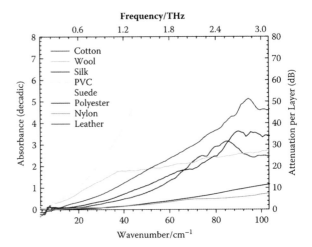

and non-destructive inspection using spectroscopy and imaging techniques. Due to these specifications, many chemical substances, pharmaceuticals and explosive materials exhibit characteristic spectral responses in this frequency range. From a pharmaceutical point of view, the average powers generated by the pulsed technique do not induce chemical or phase changes in the material under investigation. Detection and imaging of drugs, explosives and bio-agents; non-destructive and non-invasive testing and inspection, pharmaceutical and foodstuff quality control, and medical diagnostics are other applications related to terahertz spectroscopy.

Recently, the idea of developing a material catalog for terahertz region has been proposed. Substances as diverse as plastic explosives, aspirin, and amphetamines have significant absorption and reflection features in the THz range as can be seen from Fig. 1.6. The range of potential applications is likely to grow even further with the increased availability of absorption spectra (i.e. fingerprint spectra).

1.3.4 Industrial

Perhaps high cost of the terahertz technology is the main obstacle for industrial improvement of the terahertz measurement instruments. THz radiation goes through many everyday objects that don't contain water or metals. Water and metals effectively absorb/reflect THz radiation. THz radiation does, however, penetrate cardboard, paper, dry wood, various paints, many plastics and many ceramic materials. As an industrial application, characterization of the water content of newspaper print has been reported [19]. Optically-pumped far-IR lasers can also detect small voids and recognize their size and position inside electric power cables [20].

1.3 Terahertz Applications and Opportunities

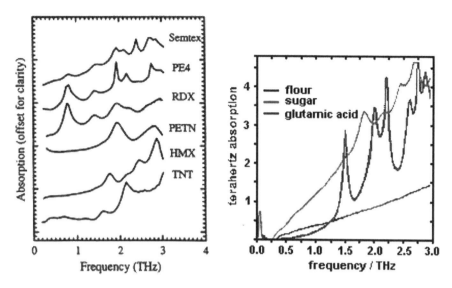

Fig. 1.6 Absorption spectra of different explosives (*left*) and other substances (*right*) measured at THz frequencies (*Courtesy* of Teraview Ltd.)

1.3.5 Communication

Although the atmosphere is almost opaque for terahertz radiations, but some close-in systems have been proposed for terahertz communications [21]. Using terahertz carriers in secure communications has lead to some advantages namely small antenna sizes and large information bandwidth. Lower scattering and much greater penetration through aerosols and clouds have made terahertz signals more favorable compared with IR and optical wavelengths for communication in the stratosphere (air-to-air links). In the presented scheme of Fig. 1.7 the bandwidth and number of discrete channels have increased and the capability of tracking and scanning have added to a terahertz communications system due to utilizing optical keying techniques and terahertz power generation via photomixer arrays [22]. In this communication system, airways data rates in the tens of gigabits per second are achievable.

1.3.6 Simulating the Radar Scattering Signatures

Due to the small spot size of terahertz wavelengths, terahertz sources can be used to illuminate scale models of large objects. Therefore, the radar scattering signatures (RCSs) obtained at much lower frequencies on actual equipment such as planes, tanks, and battleships, can be processed with a terahertz based system [23, 24].

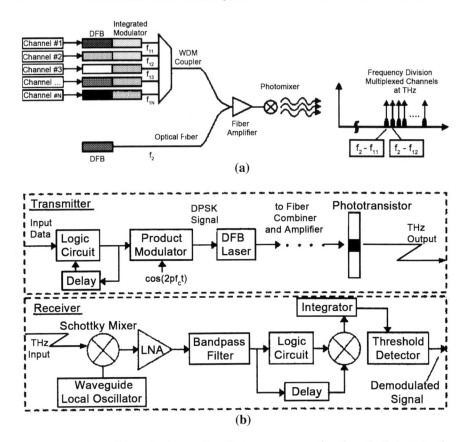

Fig. 1.7 a 1.5-μm fiber technology and terahertz power generation through photomixing for optically diplexed terahertz transmitter source. **b** Gigabit data-rate transmit/receive module based on this technology [22]

1.3.7 T-Ray Imaging

Imaging at THz frequencies of the electromagnetic spectrum has progressed rapidly in the past few years. The terahertz attentions in the area of imaging are due to substantially the following reasons: Firstly, terahertz radiation is not ionizing and it only triggers vibrational and rotational states of molecules while leaving the electronic state unaltered. Hence, terahertz radiation is inherently safer than X-rays. On the other hand, terahertz radiations are capable to penetrate a wide variety of non-conducting materials. Many materials such as plastics, packaging material like cardboard, fabrics and human/animal tissue are transparent for terahertz radiation and these materials leave spectral fingerprints when broadband radiation passes through. The much larger wavelength in comparison with wavelengths of the visible spectrum also dramatically suppresses Rayleigh scattering. There exists a significant collection of data regarding transmission,

1.3 Terahertz Applications and Opportunities

absorption and reflection of different materials. In general, the transmission decreases with increasing frequency. However in many cases it is sufficient for remote detection of hidden objects. A technical advantage is that THz waves, like light waves, can easily be propagated though space, reflected, focused and refracted using THz optics. The short wavelength, much shorter than that of microwaves, allows for a spatial resolution, which is sufficient in many imaging applications. Detection of concealed weapons, hidden explosives and land mines, improved medical imaging and more productive study of cell dynamics and genes, real-time "fingerprinting" of chemical and biological terror materials in envelopes, packages or air (security inspection) and better characterization of semiconductors (quality control) are of the other applications of terahertz radiations in imaging (Figs. 1.8, 1.9).

The terahertz time domain spectroscopy or T-ray imaging is a technique for measuring and revealing spectral content, refractive index determination, amplitude and phase, sample thickness and direct signal strength imaging based on transmitted or reflected terahertz energy from a considered sample [26]. For instance, Fig. 1.10 schematically presents a real time imaging system. The operation principles involve generating and then detecting terahertz electro-magnetic transients that are produced in a photoconductor or a crystal by intense femtosecond optical laser pulses. In a photoconductive generator, two electrodes that make up an antenna are deposited on top of a fast photoconductive material. The arms of the antenna are biased with respect to each other. When the photon energy exceeds the bandgap, electron–hole pairs are created and the free carriers then accelerate in the static field applied across the electrodes to form a transient photocurrent. If the lifetime of the photogenerated charge carriers is sufficiently short, the fast, time varying current radiates electromagnetic wave pulses. Most of

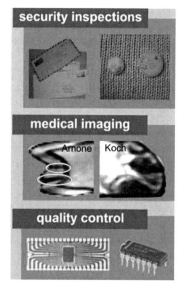

Fig. 1.8 Applications of terahertz imaging in security inspections, medical diagnosis and industrial quality control [25]

Fig. 1.9 THz image of a weapon hidden in a paper box (*Courtesy*: M. Koch, TU Braunschweig)

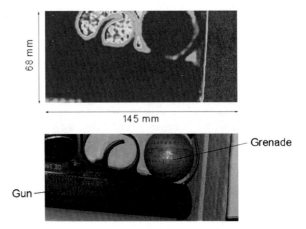

Fig. 1.10 Real time imaging systems [25]

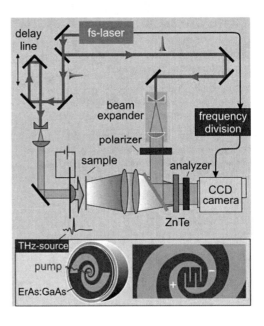

the terahertz radiation is emitted into the substrate and is collected with an appropriate lens mounted on the back of the substrate. The radiation can be detected after transmission through the sample with an electro-optic crystal, which exhibits birefringence as a result of the incident terahertz radiation. A second time delayed and polarized laser beam will then undergo a polarization change which depends on the size of terahertz electric field at the time of incidence. This polarization change is easily detected with an analyzer. A full image of the sample is obtained with the help of a CCD camera. By tuning the time delay of the second laser beam, the full time dependent terahertz signal can be recovered both in magnitude as well as phase.

1.3.8 Medical

Cancers have been of critical aspects of scientific medical researches and have attracted vast attentions for novel detection methods in recent years. Breast cancer is the second most common type of cancer after lung cancer and the fifth most common cause of cancer death. Early detection of breast cancer allows treatment at an earlier stage and significantly reduces associated mortality. The current popular screening methods are self and clinical breast exams, X-ray mammography, breast magnetic resonance imaging (MRI) and ultrasonography. Among the technical screening methods, mammography uses X-rays, which afford good spatial resolution owing to the short wavelength, but are highly ionizing, posing a radiation hazard. MRI does not use ionizing radiation, however, it requires superconductors, and thus cryogenic temperatures, and is very bulky and expensive. Ultrasonography is cheap, however its resolution is poor, and it suffers from a poor signal–noise ratio. In recent years, microwave (up to 300 GHz) and THz (0.3–10 THz) frequency electromagnetic waves have received extensive attention as powerful tools for the early detection of cancer, since they provide dielectric contrast for imaging, yet are not ionizing.

Comparing these two frequency ranges, one can see that THz offers better spatial resolution in imaging due to its shorter wavelength (detection resolution being limited by diffraction). Furthermore, rich spectroscopic information can be collected as fingerprints of biological tissues since the vibrational and rotational transition energies of the biomolecular constituents of tissue fall in the THz frequency range (e.g. DNA and proteins [27]). Since water absorbs THz waves strongly, in vivo usage of THz imaging has been primarily on skin, dentistry and breast cancer. The in vitro imaging and spectroscopic studies using THz waves are applicable to a wide range of materials. For example, images on Fig. 1.11 show an optical image (color) and a THz image (B&W) of an esophageal cancer with the circles marking the metastasis (http://www.pi1.physik.uni-stuttgart.de/research/Methoden/THzMikroSpektrometer_e.php).

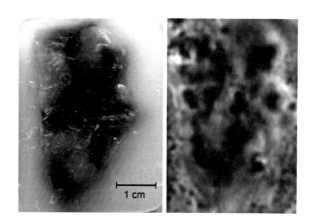

Fig. 1.11 Optical image (*left*) and a THz image (*right*) of an esophageal cancer

The structure and also the chemical content of a tissue can influence the water absorption mechanism in THz imaging. THz imaging has proposed novel features in diagnosing different deficiencies in teeth, skin, breast and solid organs which are not apparent with other imaging methods.

Determination of the extent and depth of a basal cell carcinoma tumor through reflectance mode imaging is one of the THz applications. This capability is illustrated in Fig. 1.12 where it is difficult to determine the extent or depth of this locally invasive tumor (part (a)). However, it is possible to diagnose surface details (part (b)) and the depth of the tumor (part (c)) using broad-band reflected THz signals obtained with a photoconductive source and detector. The obtained images are correlated with standard pathology where the tissue appears colored due to staining with Hematoxylin and eosin to identify pathologic characteristics [28].

In addition to imaging features and sub-surface on skin, it is possible to detect dental caries that are not evident in X-ray images and also 3-dimensional imaging of bone, although not in vivo. Tooth decay or cavities (dental caries) can be detected through absorption measurement of THz signals. Since caries are simply cavities or holes in the tooth, the caries can be detected when a T-ray is sent in (Fig. 1.13). The presence of dentine and enamel, or only dentine, can also be identified by measuring the difference of refractive index over the surface of a tooth.

1.3.9 Physics [28]

By analyzing the THz spectrum, several interesting features have been apparent to be in this region such as rotational excitations in molecules, Rydberg transitions in

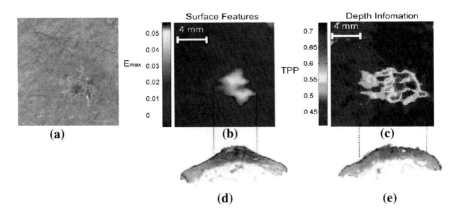

Fig. 1.12 An in vivo measurement of a basal cell carcinoma. **a** A clinical photograph of lesion, **b** surface features detection through THz imaging, **c** tumor extent measurement, **d** histology section showing acute inflammatory crust corresponding to THz image **b** and **e** histology section showing lateral extent of tumor corresponding to THz image **c** [28, *Courtesy* of TeraView Ltd, Cambridge, UK]

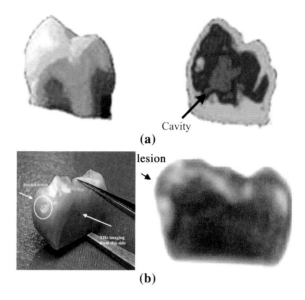

Fig. 1.13 a Detection of dental caries through THz 3-dimensional imaging and **b** photograph (*left*) and near-field THz image (*right*) of a human tooth [28, *Courtesy* of D. Fried, U.C. San Francisco, and U. Schade and K. Holldack, BESSY Gmbh]

all Coulomb bound systems and excitons in solids. On the other hand, highest degree of control and characterization of the electromagnetic field on a sub wavelength are achievable in THz spectrum owed to advances in ultrafast optics. THz is the highest frequency band where the field can be measured coherently without an interferometer.

Anderson localization [29] is one of the interesting optical physics phenomena in random media which have seen at microwave frequencies so far. Observation of such effect in visible or near-infrared encounters a difficulty to specify that this effect is not only due to residual absorption in the dielectric. However, this localization can be observed easily at THz frequencies due to higher dielectric contrast and virtually zero absorption. This can be promising in development of random lasers [30]. The possibility of studying multiply scattered waves and wave diffusion phenomena in THz region can also be useful in some application fields such as medical optics and geophysics.

Beside the applications of terahertz in physics, there are several potentially interesting opportunities in THz domain. Artificially structures electromagnetic meta-materials could exhibit a negative permittivity and a negative permeability at microwave frequencies which make them distinct from the constituent materials. It is reported that these materials exhibit large losses at higher frequencies. There is a clear desire to develop the application range of these materials to higher spectral ranges and the THz region is the most demanded region.

Terahertz techniques can be used to measure and study of the relationship between the magnetic near and far fields as well as the transition region in between. Therefore, a better understanding of near filed electromagnetic measurements can be obtained.

Plasmas created by high-intensity femtosecond lasers can serve as a source of intense terahertz emission with possible THz pulse energies approaching a fraction of a mJ within an ultra-short pulse.

1.3.10 Chemistry and Biology [28]

The control of chemical reaction in chemistry and biology is a great challenge and it is necessary to channel the energy inserted to a molecular system into specific mode to control the chemical reactions with high accuracy. It might be useful to notify that the chemistry and biology both are common in the sense of dealing with molecular interactions. However, chemistry deals with small molecules and repetitive polymers while the biology is considered to deal with complex macromolecules, cells and tissues. As it is shown in Fig. 1.14 there are two extreme for motion of a general biological molecule: the first extreme is the vibrational modes localized on particular functional groups in amino acids which occur on femtoseconds to tens of femtoseconds of time scales and are in infrared region of the spectrum; the second extreme is the tertiary structural dynamics which occur on time scales of nanoseconds to milliseconds. Low frequency intra and intermolecular modes such as secondary structural modes that reside between these two extreme, occur on a picoseconds to tens of picoseconds time scale and fall within the THz region of the spectrum.

In the case of biological systems, it is known that these systems consist of time-varying complexes of macromolecules. The local structure of a biological complex, such as a single cell changes with time depending on the state of the cell cycle and there are a few methods to measure these chemical changes. THz region

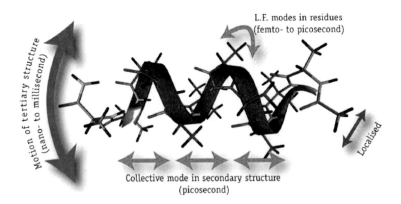

Fig. 1.14 Schematically presentation of a general biological molecule with related dynamical time scales [28, *Courtesy* of K. Wynne, University of Strathclyde]

can be suitable to imagine the changes in complex molecular interactions since it is sensitive to changes in the collective modes of a system.

An important feature of biological molecules is their relation with water. For example, protein molecules must be in aqueous environment to function. Since water at the air/water, water/solid and water/membrane interface of biology has different characterization such as density, pH, orientational order, hydrogen-bonding, etc., spectroscopy in THz region should present different signatures of these biological systems compared with bulk liquid water.

Large amplitude THz fields (as large as 0.1–1 MV/cm) can be used to control electron transfer reactions in chemistry and biology. High-amplitude fields can change the transfer rate and therefore highly influence the photochemical electron transfer yield. THz can effect on energy and dynamic characteristics of chemical interactions since its amplitude could be comparable to the electrochemical potential difference that drives the electron transfer reaction. Also, THz fields with large amplitudes can be applied toward drive and control direction selective electron and wrong-way electron transfer as shown in Fig. 1.15.

THz spectral region also probes the rotational dynamics of light molecules, the low frequency bending and torsional modes of heavier molecules and collective intermolecular modes of molecular clusters.

The phonon modes of inorganic and organic crystals fall in the THz region of the spectrum. Equilibrium measurements as well as dynamical processes can be probed.

THz spectroscopy of molecular complex in the cell will provide an important window into label-free measurements of protein–protein interactions. This feature could be interesting in cellular imaging. Achievement to AFM-like tips with probe region selectivity of about tens of nm is one of the important technological challenges toward the cellular imaging field.

There are also other opportunities for THz radiations such as nonlinear spectroscopy and electron spin resonance (ESR) spectroscopy studies in chemical and biological systems.

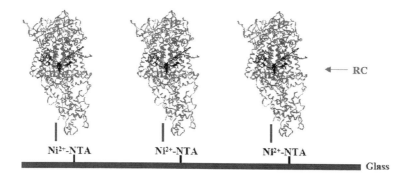

Fig. 1.15 Oriented bacterial reaction centers [28, *Courtesy* of K.V. Lakshmi, Yale University]

In the following sections, we will briefly highlight specific components that are employed for the applications mentioned here and some of the instrumentation that has been constructed from these components for measurement and test.

1.4 Terahertz Components

Here, we introduce a few of the essential component technologies that have been developed for terahertz applications. These components can be divided into two main categories: detectors and sources.

1.4.1 THz Detectors

Today, terahertz detectors as the pioneer of submillimeter-wave technology are developed to measure broad and narrow band signals in a wide range. Detection at terahertz frequencies obviously differs from detection at shorter optical wavelengths and longer radio wavelengths. In comparison to shorter wavelengths, due to the low level of photon energies at terahertz frequencies, ambient background thermal noise is dominant that appeals cryogenic cooling or long-integration-time radiometric techniques or both. Besides Airy disk diameter is rather large that makes a mode converter or matched director (antenna) between the signal and detector element necessary. It is true to say that in terahertz field, RF detection shows better performance than optical detection because the crossover frequency at which an ideal thermal noise limited detector (such as a room-temperature Schottky barrier diode) excels the sensitivity of an ideal quantum detector (like a photodiode), falls between 1 and 10 THz [31]. In comparison to longer wavelength radio techniques, lumped electronic components such as resistors, capacitors, and inductors, as well as amplifiers and low-loss transmission media are not available to terahertz detectors. Heterodyne detectors are the most common terahertz detectors although nowadays focuses are toward direct detection techniques and components. The following discusses heterodyne and direct detectors technologies that can be realized in room-temperature and cryogenic formats.

Heterodyne Semiconductor [3]: Semiconductor detectors have an adequate level of sensitivity that is suitable for Earth science and planetary applications. Heterodyning reduces bandwidth and thereby signal-to-noise is increased. The present low-noise amplifiers operate up to 150 GHz and therefore frequency down-conversion (crystal rectification) and post or IF amplification is applied [32]. Radiometric techniques are still employed for acquiring weak signals embedded in background noise.

For applications where the sensitivity of room-temperature detectors is adequate, the basic single Schottky diode mixer is the preferred down-converter in the terahertz frequency range. For example, planar diode mixers (Fig. 1.16) are space

1.4 Terahertz Components

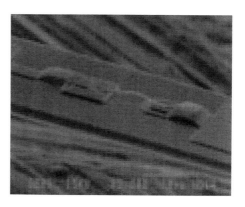

Fig. 1.16 Planar integrated submicrometer area Schottky diode on 3-μm-thick 2 30-μm-wide GaAs membrane for operation at 2500 GHz, also coupled to a waveguide mount [33]

qualified for frequencies as high as 2500 GHz with noise performance below 5000-K double sideband [33]. The main problem for this configuration is high LO power since a semiconductor down-converters which work based on nonlinear effect require ≈ 0.5 mW of RF drive level at frequencies close to observed signal frequencies. Receivers based on room-temperature Schottky diode mixers typically have radiometric sensitivities 0.5 K at 2500 GHz and a 1-GHz predetection bandwidth which is enough for detecting many naturally occurring thermal emission lines. Cooling of the detector can improve receiver performance [34].

Heterodyne Superconductor [3]: Cryogenic cooling is inevitable to obtain high-sensitivity in the terahertz detectors. Detectors based on the Josephson-effect, superconductor–semiconductor barriers, bolometric devices, the superconductor–insulator–superconductor (SIS) tunnel junction mixer, and are different types of superconducting heterodyne [3, 35].

In fact SIS is the superconducting equivalent of the Schottky diode down-converter. The current flow mechanism is based on the photon-assisted tunneling process. SIS mixers can operates up to 1200 GHz [36]. The SIS mixer uses nonlinear I–V characteristic created by the sharp onset of tunneling between the single-electron quasi-particles on either side of a thin superconducting gap (Fig. 1.17). In view of the fact that the tunneling process is a quantum mechanical phenomenon, sensitivity of the SIS mixer is restricted to the quantum limit (factor of ten of $T_m = h\upsilon/2k \approx 0.05$K/GHz for frequencies up to 1 THz: T_m is the equivalent mixer noise temperature which is governed by the Heisenberg uncertainty principle). The LO power in the SIS mixers is on the order of microwatts and this is a noticeable advantage. SIS devices reach a natural frequency limit that depends upon the tunnel junction material composition ($f_{cutoff} \approx 146 T_c$ where T_c is the critical temperature).

Hot electron bolometer (HEB) mixer is an alternative to terahertz SIS mixers [37]. Modern HEB mixers are generally based on micro-bridges of aluminum–nitride that respond thermally to terahertz radiation (Fig. 1.18) [38]. Due to fast phonon or electron diffusion cooling process, HEBs have very high speeds. Voltage responsivity at picoseconds range allows the bolometer to track the IF up to several gigahertz in heterodyning systems. The operating and material critical

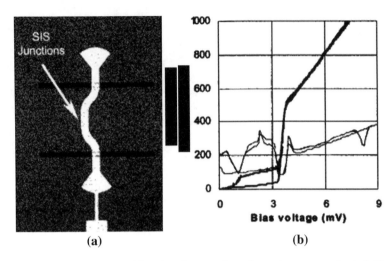

Fig. 1.17 a 1200-GHz SIS tunnel junctions integrated on planar slot antenna circuit. **b** Current–voltage relationship and IF output signal for 77- and 300-K blackbody loads. Equivalent input receiver noise temperature is 650-K DSB at 1130 GHz [36]

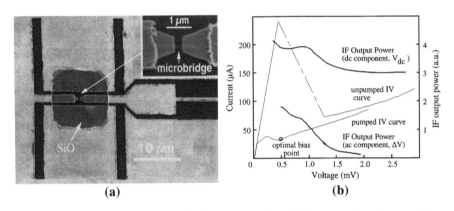

Fig. 1.18 a 2.5-THz niobium HEB twin-slot antenna mixer. **b** Current–voltage characteristics with and without LO power applied. Also shown is the dc (Vdc) and ac (ΔV) components of the IF output power [40, 41]

temperatures determine the HEB noise performance. LO power in these systems is low (nW range) [39].

Direct Detectors: For applications that do not require ultrahigh spectral resolution, direct detectors are being replaced to heterodyne systems. GaAs-Schottky diodes (used as antenna-coupled square-law detectors), conventional bolometers (based on direct thermal absorption and change of resistivity), composite

1.4 Terahertz Components

bolometers (have the thermometer or readout integrated with the radiation absorber), microbolometers (that use an antenna to couple power to a small thermally absorbing region), Golay cells (based on thermal absorption in a gas-filled chamber and a detected change in volume via a displaced mirror in an optical amplifier), an acoustic bolometer (which reads out the change in pressure of a heated air cell using a photoacoustic detector) and a fast calorimeter (based on single-mode heating of an absorber filled cavity) are the THz direct detectors [42, 43]. The antenna-coupled detectors suffer from calibration problem and the problem with acoustic bolometer and fast calorimeter is the response time that is on the order of seconds. Helium-cooled silicon, germanium, or InSb composite bolometers are the most common cooled detectors which have response times on the microsecond scale. Some traditional IR detectors respond in THz range. These detectors are pyroelectric (that changes its dielectric constant as a function of temperature) and direct photoconductors based on mechanically stressed gallium-doped germanium or even HgCdTe [44]. Different forms of cooled bolometers of many forms such as the transition edge bolometers are being used for THz-spectroscopy [45, 46].

In the recent years a new detector has been proposed. This detector is the quantum-dot single-photon detector which uses a cold single electron transistor (SET) and quantum dot in a high magnetic field (Fig. 1.19) [47, 48]. Incident terahertz photons are coupled into the quantum dot via small dipole antennas. In the quantum dot, incident photon creates an electron–hole pair which causes a polarization between two closely coupled electron reservoirs. Electron tunneling occurs, causing a shift in the gate voltage of the SET. As the results, this detector

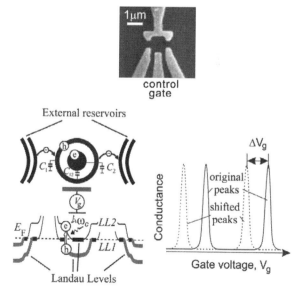

Fig. 1.19 Komiyama quantum-dot single-photon detector concept [50]

operates in the range of 1.4–1.7 THz and has NEP is on the order of 10^{-22} W/$\sqrt{Hz}$, more than 1000 times more sensitive than the best bolometric devices. The weak point of this quantum dot detector is its speed (around 1 ms) that can be tackled by the new RF SET devices (have intrinsic speeds near 10 GHz) [49].

In the following, some techniques will be discussed in detail.

1.4.1.1 Pyroelectric Detectors

For terahertz applications that do not require high sensitivity, pyroelectric detectors have been offered. Ferroelectric materials such as TGS or Lithium Tantalate, exhibit a large spontaneous electrical polarization which has varies with temperature.

If electrodes shown in Fig. 1.20 are placed on opposite faces of a thin slice of the material to form a capacitor, an electrical signal will be obtained and a voltage across the capacitor for a high external impedance is created. These types of detectors are only sensitive to AC signals (i.e. time-varying) and the capability of room temperature operation and small detector area size (which can give fast thermal response time (<1 µs)) are other features of these devices. Pyroelectric effect is sensitive only to heat not to wavelength and wavelength selection is made through selection of appropriate window material.

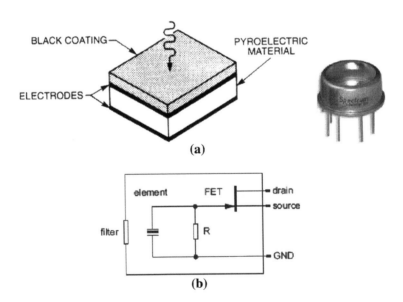

Fig. 1.20 **a** structure of a pyroelectric detector and **b** operational circuit diagram of detector. A low noise JFET at the input stage ensures that the amplifier noise contribution does not dominate the system noise performance [51]

1.4.1.2 Gallium Doped Germanium Photoconductive Detector

Gallium doped Germanium (Ge:Ga) has photoconductive response out to wavelengths longer than any other combination of elements. In unstressed configurations, the cut-off wavelength is approximately 120 um. By applying stress to the detector crystal, this can be extended to beyond 200 um. The measured spectral responsivity of stressed and unstressed detectors is shown in the Fig. 1.21.

The performance characteristics of the detector depend critically on the operating conditions. In low radiation background conditions such as satellite astronomy, detector NEP can exceed 1 fW $Hz^{-1/2}$, though the speed of response will be limited, perhaps to 20 Hz (-3 dB). In higher background laboratory applications the detector may operate at 50 kHz (-3 dB), albeit with much reduced sensitivity.

1.4.1.3 Bolometer Detectors

In 1878 Samuel Pierpont Langley invented the bolometer, a radiant-heat detector that is sensitive to differences in temperature of one hundred thousandth of a degree Celsius (0.00001°C). This instrument enabled him to study solar irradiance far into its infrared region and to measure the intensity of solar radiation at various wavelengths and is composed of two thin strips of metal, a Wheatstone bridge, a battery, and an electrical current measuring device [53]. A bolometer is a type of thermal detector where the electrical resistance of the material is the property that is measured in response to incident electromagnetic radiation especially in the far infrared/mm wave region. The active element may be a metal, a semiconductor and a superconductor. Thermal detectors convert incident radiation to heat and the active element responds to the heat input which causes a change in some measureable physical property of the device. Composite bolometers consist of a number of components, each performing a particular function (Fig. 1.22). Absorption of input radiation occurs in a thin impedance-matched film.

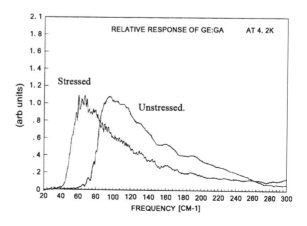

Fig. 1.21 Photoconductive response of stressed and unstressed Gallium doped Germanium at 4.2 k [52]

Fig. 1.22 Structure of a bolometer detector [51]

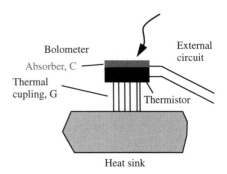

The thermistor, which is in direct thermal contact with the film, can therefore be much smaller and have reduced thermal capacitance. This in turn leads to a shorter thermal time constant.

1.4.1.4 Composite Germanium Bolometer

In QMC Instruments Ltd a composite bolometer has designed for operation at cryogenic temperatures near 4.2 K and incorporates a Germanium thermistor attached to a metallic absorber which is in turn deposited onto a thin silicon nitride (as an isolator) support substrate [52]. Absorption of radiations through the metal film, increases the temperature of the film and hence the thermistor. This phenomenon changes the electrical impedance of thermistor and this change is sensed as a voltage change at the input of the amplifier (Fig. 1.23).

The bolometer usually mounts in an optical integrating cavity behind Winston Cone coupling optics along with low-pass filters. This filter efficiently rejects the unfavorable high frequencies. The field of view of the coupling optics should be matched to the incoming beam geometry to ensure that all available signal power is available for detection and also that the bolometer is not unduly de-sensitized by exposure to an unnecessarily high background power.

1.4.1.5 Untuned Indium Antimonide

The bolometer based systems suffer from low response speed which is limited by the thermal capacitance of the thermistor and even in a composite structure this speed reaches to order of 1 ms. This deficiency limits the usage of these detectors in high-time-resolution applications. Indium Antimonide (InSb) is commonly used as a near-infrared detector at liquid nitrogen temperature 77 K. At 4.2 K, free-electron absorption of rather longer wavelengths results in changes to the electron mobility. Since the electron–phonon relaxation time—by which energy is lost to the lattice—is highly temperature dependent and is relatively high (~ 300 ns at 4.2°k), the much faster electron–electron interaction time brings the carriers into

1.4 Terahertz Components

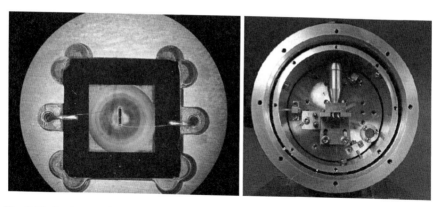

Fig. 1.23 Implementation of a composite germanium bolometer [52]

Fig. 1.24 Implementation of a InSb hot-electron bolometer [52]

thermal equilibrium at a temperature above the lattice. This modification in electrons equilibrium temperature changes electron mobility and is known as the hot-electron response.

Figure 1.24 presents a high sensitivity InSb hot-electron bolometer from QMC Instruments Ltd. that can be operated at or below 4.2K where the InSb (size 5 mm × 4.7 mm × 300 μm) detector has mounted on a quartz substrate.

The sensitivity of InSb hot-electron bolometer at frequencies between 500 GHz and 2.5 THz can be improved using magnetic resonance. In InSb at 4.2 K the electron–electron interaction time is 300 fs and this determines that absorption will be constant up to 500 GHz, beyond which it falls as the signal frequency to the power -1.8. Permanent magnets can be used to apply the magnetic field to detector. The maximum magnetic field strength applicable to the detector through permanent magnets limits the maximum operation frequency up to 2.5 THz.

1.4.1.6 Golay Cell Detectors

A Golay cell detector is a room temperature bolometer, which is a convenient choice for the moderate to high intensity THz signal measurements. This detector is based on the volume or pressure change (thermal expansion) of an encapsulated gas with temperature (Fig. 1.25). The volume change is measured for example by the deflection of light-rays resulting from the motion of properly positioned mirrors fastened to the walls of the gas container. Cell is a metal cylinder having a blackened metal plate at one end and a flexible metalized diaphragm at the other. It is filled with an inert gas and then sealed. Radiation incident upon the blackened metal plate is absorbed and heats the gas which increases the pressure thereby deforms the diaphragm. Light is reflected by the diaphragm motion onto a photocell to measure the incident flux. Wide spectral range of $\sim$20 GHz–20 THz can be detected using Golay cell detectors.

1.4.1.7 Terahertz Electronic Component [54]

Terahertz electronics technologies have potential to replace or augment more traditional terahertz photonics techniques, such as Time Domain Spectroscopy, Fourier Transform Infrared Spectroscopy and THz dielectric spectroscopy [55–58]. Beside the THz electronics technologies relying on the frequency multiplication mostly using Schottky, Gunn and IMPATT diodes, THz transistor technology has proposed, since the device feature sizes have condensed to the point, where ballistic mode of electron transport becomes dominant [59, 60].

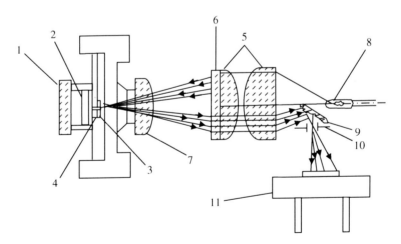

Fig. 1.25 Golay cell detector setup. A modulated signal (*1*) is incident upon the semi-transparent film (*2*). This heats the gas in the chamber and distorts the membrane forming the all of the chamber. An LED (*8*) sends a signal onto the mirrored back surface of the chamber containing the absorbing membrane. The modulated signal appears as a modulated optical signal [51]

1.4 Terahertz Components

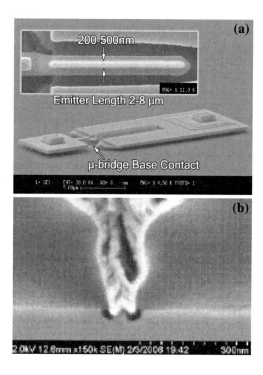

Fig. 1.26 a SEM image ultra-high frequency HBT and **b** Cross section of 35 nm InP HEMT T-gate from Northrop–Grumman [54, 63, 64]

Silicon transistor, Heterostructure Bipolar Transistors (HBTs) and High Electron Mobility Transistors (HEMTs) are capable of operation in the sub-terahertz region and recently have reached cutoff frequencies approaching 1 THz [61, 62]. Researchers have recently reported on GaAsSb/InP-HBT with f_T = 670 GHz, InP-HEMT with f_{max} greater than 1 THz, MMIC amplifier operating up to 600–700 GHz and RF performance of silicon SOI achieving the cutoff frequency 485 GHz with a 45 nm NMOS transistor (Figs. 1.26, 1.27) [63–65].

The new THz electronics relies on using wave of electron density (called plasma oscillations) for detection and generation of THz radiation. This approach was recently used to demonstrate THz detection by Si CMOS (Fig. 1.28) [66–69]. THz emitters based on the excitation of two-dimensional electron gas (2DEG) plasmons at semiconductor hetero-interfaces have been also demonstrated. These sources are tunable and can be used together with THz plasmonic resonant and non-resonant detectors using the same technology.

1.4.1.8 Traveling Wave Terahertz Detector

It has been proposed to use antenna coupled traveling-wave diode detectors to achieve high performance in the infrared. These devices channel radiation received by an antenna into a parallel plate waveguide: a metal–insulator-metal (MIM)

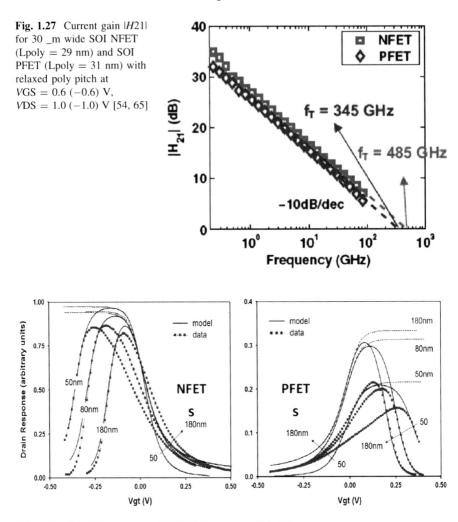

Fig. 1.27 Current gain $|H21|$ for 30 _m wide SOI NFET (Lpoly = 29 nm) and SOI PFET (Lpoly = 31 nm) with relaxed poly pitch at $VGS = 0.6\ (-0.6)$ V, $VDS = 1.0\ (-1.0)$ V [54, 65]

Fig. 1.28 Sub-THz response of CMOS Si detectors [54, 70]

diode that rectifiers the carrier, producing a signal voltage. This arrangement facilitates a better impedance match between the antenna and the rectifying MIM diode than a lumped-element diode connected to the antenna.

The antenna and rectifier pair is an extremely efficient means for detecting and converting electromagnetic energy. Since it does not depend on electron transitions between different energy states, it does not suffer from the thermally-imposed limitations of conventional interband or impurity-band based detectors, and has been shown to operate at room temperature. The challenge in detecting THz or higher frequency waves lies in the antenna design and the operation speed of the rectifying diodes. An example of such a detector which is designed to receive radiation in the wavelength range of 9–12 μm is shown in Fig. 1.29.

1.4 Terahertz Components

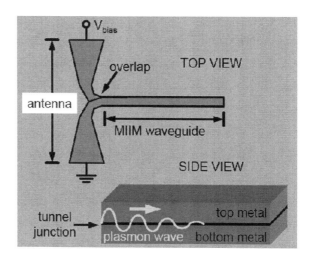

Fig. 1.29 Traveling wave terahertz detector [71]

The MIM diode works based on the quantum tunneling of electrons (travel between metal electrodes without being excited over the insulators band gap barrier). In the MIM diode coupled to an antenna, the incident photons excite surface plasmons on the metal antenna. These surface plasmons, due to the resonant geometry of the antenna, transfer and focus the electric field of the incident photons across the insulator of the MIM diode. As a result, the electric field is directed across the insulator of the MIM diode and the quantum tunneling of electrons can be accomplished. This allows MIM diodes exhibit the non linear I(V) characteristic required for signal rectification: the conversion of an alternating current (AC) signal into a direct current (DC) level. Tunneling characteristics vary with shape and size of the insulator barrier [72]. A bow-tie antenna is selected for its wider bandwidth and grater response where the arm length is designed to be 6 μm which respond to wavelength of 10.6 μm. Recently the Metal-Oxide-Metal (**MOM**) based THz detector by using nanoscale metallic antenna structures has developed that can operate at >33 THz (∼10 μm) [73]. Figure 1.30 shows the SEM image of a MOM-based detector and its polarization-sensitive detection of ∼10 μm radiation.

1.4.1.9 Tunable Plasma Wave-HEMT THz Detector

Conventional solid-state all-electronic THz detectors such as bolometers, Schottky diodes, pyroelectric or photoconductive devices have a broadband performance and require cryogenic operation temperatures. However, the Tunable Plasma Wave (TPW)-based devices benefit from all-electronic tunability owned to Plasmon resonances in a gated 2-dimensional electron gas (2DEG). In this device the incident THz waves excite the plasma waves in the 2DEG and the resonant voltage response to THz radiation can be tuned to various THz frequencies by changing

Fig. 1.30 a SEM image of MOM-based THz detector and b its polarization-sensitive detection [73]

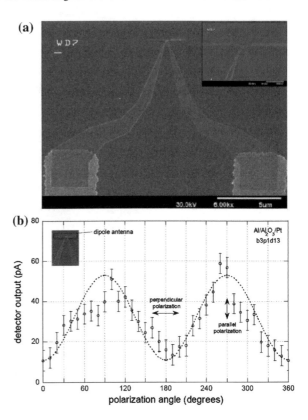

the electron density in the 2DEG [74]. In AlN/GaN heterojunctions tunable plasmon resonance frequency can be extended up to 100 THz due to their extraordinarily high carrier concentration with high carrier mobility. Figure 1.31 shows 2DEG carrier concentrations and mobilities measured on AlN/GaN structures in comparison with other 2DEG systems [75].

An SEM image of such a HEMT-based detector is shown Fig. 1.32. Since the plasmon resonant frequency is proportional to the square root of the electron concentration, so the plasma frequency in a bulk metal is generally in the ultraviolet rang, while the 2DEG formed at an AlGaAs/GaAs HEMT corresponds to a plasmon resonant frequency up to ~ 10 THz. The frequency range in AlN/GaN interface is higher (~ 3 times) than AlGaAs/GaAs interface due to the higher 2DEG density in AlN/GaN interface which can be an order of magnitude larger than AlGaAs/GaAs junction. Carrier mobility is also very important since resonant detection requires coherence of the plasma wave carried by the 2DEG carriers: the higher the mobility the longer the plasma wave will maintain its coherence, thus the more sensitive is the detector. Figure 1.33 shows the calculated plasma wave frequencies of 2DEGs formed in various semiconductors.

Fig. 1.31 Measured 2DEG carrier concentrations and mobilities in different structures [76]

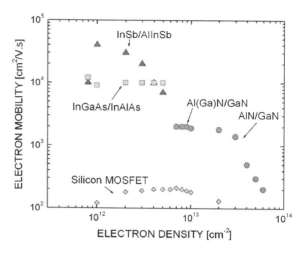

Fig. 1.32 SEM image of HEMT-based detector [76]

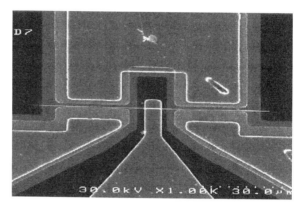

1.4.1.10 Terahertz Detectors on a Single Chip [77]

In recent years, it has developed compound semiconductor-based terahertz detectors that mainly focus on the implementation of frequency-tunable terahertz detectors based on nanostructured semiconductor systems and can be integrated seamlessly with conventional electronics. The operation principles of these detectors are based on plasmons in confined geometries such as quantum dots, and photoexcitation of electrons in quantum point contacts (QPCs). The resulting devices are expected to be suitable for integration into large-scale arrays, allowing sophisticated temporal and spatial signal processing functions.

The electrical properties of semiconductor nanodevices may be modified in the presence of terahertz radiation. Bolometric, photonic and plasmonic detectors are proposed types of far infrared (FIR) detector. In bolometric photoconductors (which are the most common type), the heating due to absorbed FIR causes a net change in conductivity. The main deficiency of these devices is that they have no

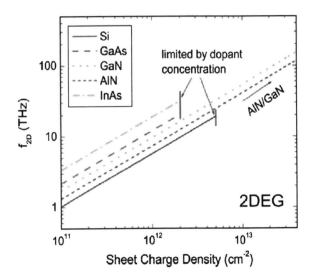

Fig. 1.33 Calculated plasma wave frequencies versus 2DEGs density in various semiconductors [76]

frequency resolution. The photonic detectors such as resonant tunneling diodes are also influenced by bolometric effects from free carrier absorption. The free carriers are strongly absorbed compared with confined two dimensional electron gas since the confined carriers only response to perpendicular component of electromagnetic field. Further confinement may be added through introducing quantum point contacts which is realized by negatively biased metal contacts (Fig. 1.34).

The quantum point contact allows the high sensitivity and the maximal photonic transitions of confinement in the plane. The QPC detector sensitivity will be reduced at higher temperatures due to thermal background dark current.

Two-dimensional and three-dimensional confinement can be used for realization of plasmonic devices and through gate defined length scales which changes the confined Plasmon resonances, we will have tunable frequency resolution. Hybrid' devices combining bolometric and plasmonic techniques are also being considered. One may replace simple gate structure of the QPC with a metal grating which creates a grating for the electron waves that is tuned to a particular terahertz frequency.

1.4.1.11 Quantum Dot Photodetector for THz Detection [79]

Detection of a broad range of infrared and THz frequencies can be implemented using quantum dot intersublevel photodetector (QDIP). Due to three-dimensional carrier confinement in quantum dots, these kinds of photodetectors take advantages of: (i) intrinsic sensitivity to normal-incidence light, (ii) long lifetime of photo-excited electrons due to reduced electron–phonon scattering, and (iii) low dark current. Recently it has been demonstrated a multi-color QDIP with a simple quantum dot heterostructure active region. The device exhibits strong absorption peaks in the 3–13 μm (MIR) and 20–55 μm (THz) ranges with large responsivity

1.4 Terahertz Components

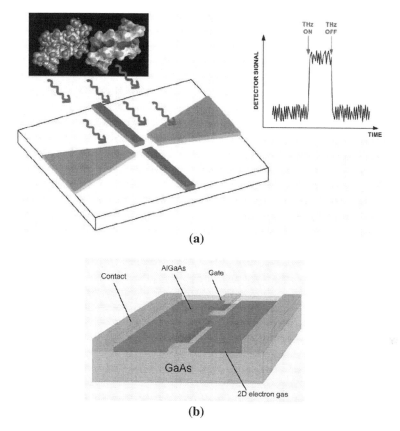

Fig. 1.34 a The *grey* and *yellow* regions create a quantum point contact nanowire device that detects terahertz radiation emitted by a targeted substance. **b** Structural schematic of the quantum point contact [77, 78]

and detectivity at temperatures up to 150 K. The multi-color characteristic of this detector originates from the existence of different electron energy states in the quantum dots. By tuning the dot size, alloy composition and barrier layer bandgap, it should be possible to extend the terahertz absorption into the 1–3 THz range. A comprehensive discussion on quantum dot THz photodetectors will be presented in Chap. 2.

1.4.1.12 Multiband THz Detection and Imaging Devices [80]

Up today, the progress in electronics conversion systems has not follow the fast development of THz sources and detection systems technology. Two main problems in THz domain are dominancy of thermal noise and material characteristics alteration by individual and collective modes of molecular vibrations and rotations. Currently, analog-to-digital (ADC) conversion systems are seen as the bottleneck for THz applications. The act of down-convert the THz frequency into GHz domain and then digitize the signal in available ADC systems not only imposes complexity into system but also increases size, weight and power requirements. Using a series of electromagnetic input resonator that selects the band and polarization in series with a plasmon resonator-amplifier the detected electromagnetic signal is applied to a very low current FET. According to the nonlinear carrier perturbation process THz signal is detected and the down-conversion to lower frequency is accomplished. The detected signal is digitized by a fast ADC and is stored in fast memories. The detection chain repeated for various frequencies and grouped in a single multiband module—eye unit, is further integrated into the imager.

1.4.1.13 Integrated Terahertz Imager Based on Quantum Dots

Distinct spectra and images in terahertz measurements require highly sensitive detectors. To obtain higher levels of sensitivity as obtained for visible light domain, two main challenges should be noticed: first, the photon energy of the terahertz wave is two to three orders of magnitude lower than that of the visible light, making the development of a high-performance terahertz detector a difficult challenge. The other problem relates to low spatial resolution of terahertz imaging originated from longer wavelengths of terahertz radiation compared with visible light. Application of nanoscale materials in recent years have been promised for dealing with these problems.

Two types of emerging terahertz detectors exploit novel nanoelectronic technologies. The first is a highly sensitive and frequency tunable terahertz detector based on a carbon nanotube (CNT) quantum dot (QD) [81] and the second type is a near-field terahertz detector for high-resolution imaging. It is found that terahertz irradiation generates new CNT currents and its peak position relative to the gate voltage linearly depends on the photon energy of the incident terahertz wave. These observations provide direct evidence of terahertz photon-assisted tunneling, demonstrating that the CNT-QDs can be used as a frequency-tunable terahertz photon detector. Also, a device for near-field terahertz imaging has developed in which all components—an aperture, a probe, and a detector—are integrated on one gallium arsenide/aluminum gallium arsenide (GaAs/AlGaAs) chip [82]. This scheme allows highly sensitive detection of the terahertz evanescent field alone, without requiring optical or mechanical alignment. The following two sub-sections review these types of detectors briefly.

1.4.1.14 CNT-Based-QD Frequency-Tunable THz Detector

Photon-assisted tunneling (PAT) is one of interesting quantum phenomenon that can be used for highly sensitive detectors. PAT is associated to electron tunneling via quantum detection of an electromagnetic wave [83]. The PAT phenomenon in

1.4 Terahertz Components

quantum dots [84–86] has attracted great attention from application point of viewpoint for a detector, the PAT in quantum dots. Since the energy level splitting in QDs based on a carbon nanotubes reaches 10 meV (corresponding to a THz frequency ∼2.4 THz), CNT-based-QDs can be suitable alternatives for THz detection.

The schematic view of the CNT-based-QD detector is illustrated in Fig. 1.35a. The structure consists of single-wall CNTs with a diameter of ∼1 nm dispersed on a GaAs/AlGaAs heterostructure wafer which is covered with ∼100 nm SiO_2 film for electrical insulation. The distance between source and drain is ∼600 nm and the whole structure was mounted in a ^{4}He cryostat at a temperature of 1.5 K, and the sample was irradiated with a THz wave through an optical window made from a Mylar sheet. This device functions at 4 K properly, which are inaccessible via conventional AlGaAs-based single electron transistors.

Figure 1.35b represents a schematic diagram of electron tunneling processes in a QD in the presence of an electromagnetic wave. Energy states in the QD can be tuned by changing the electrostatic potential with application of a gate voltage. When the Fermi level in the source is aligned with a level in the QD, a source-drain current flows via elastic tunneling as presented in the middle part. Otherwise, electrons can tunnel out inelastically through photon absorption (PAT) as depicted in the other parts of Fig. 1.35b. In the latter case, a new current is generated in the Coulomb blockade regime with irradiation of the electromagnetic wave. The peak position of the new current with respect to the gate voltage depends linearly on the photon energy (hf).

Figure 1.36 shows the source-drain current as a function of gate voltage with and without THz irradiation at a source-drain voltage of 1 meV. New satellite peaks appear in the Coulomb blockade regime in the presence of THz irradiation compared with the case that no THz irradiation exist (black curve). Also, increasing the frequency of THz wave shifts the satellite peaks toward positive direction of gate voltage. The inset of Fig. 1.36 depicts the energy spacing, $\kappa \Delta V_G$, between the original peaks and the satellite peaks as a function of the photon energy, hf, of the THz wave where κ is the conversion factor, which is defined as the conversion ratio of gate voltage into energy. This measurement denotes that the

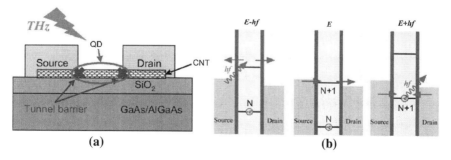

Fig. 1.35 **a** Device structure of a carbon nanotube quantum dot (CNT-QD). **b** Schematic view of electron tunneling processes in a QD in the presence of an electromagnetic wave [81]

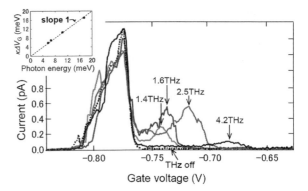

Fig. 1.36 Source-drain current as a function of gate voltage in the presence and without THz irradiation. The *inset* shows the energy spacing, $\kappa\Delta V_G$, between the original peaks and the side peaks as a function of the photon energy of the THz wave [87]

position of the satellite peaks depends linearly on the photon energy. These results provide evidence for THz-PAT in the CNT-QD. The proposed findings express that the CNT-QD THz detector can be a promising candidate for highly sensitive and frequency-tunable THz detector for THz microscopy and spectroscopy.

Since the coupling between the THz wave and the CNT-QD whose size is much smaller than the wavelength of the THz wave is not efficient, some methods may considered to improve the coupling efficiency such as: Implementing an antenna structure as the source and drain electrodes, and by placing a Si hemispherical lens on the back surface of the substrate of the CNT-QD or capacitive coupling a CNT-QD with a quantum point contact device on a GaAs/AlGaAs heterostructure (which makes it possible to observe a single electron dynamics) [88–90].

Two-dimensional arrayed configuration of CNT-QD detectors can be considered as an interesting application of this type of detectors for a THz camera for real-time THz imaging [91].

1.4.1.15 On-Chip Terahertz Detection

High spatial resolution in optical imaging can be obtained through two methods: a solid immersion lens and near-field imaging. The resolution of imaging in the case of solid immersion lens is restricted by the diffraction limit [92]. Near-field imaging technique can be considered as a powerful method to overcome the diffraction limit using a tapered, metal-coated optical fiber or a metal tip, and either a waveguide or a coaxial cable. This technique has been implemented in visible and microwave regions while the development of near-field imaging in the terahertz region has faced several difficulties such as lack of terahertz fibers or other bulk terahertz-transparent media suitable for generating near-field waves, and also low sensitivity of commonly used detectors in the terahertz region.

Measuring the propagating field raised from the scattering of the near-field (evanescent) wave has been mediated with distant detectors in conventional near-field imaging systems. This method is influenced from far-field waves whilst requires detection of very weak waves. In contrast, the introduced near-field terahertz imager

1.4 Terahertz Components

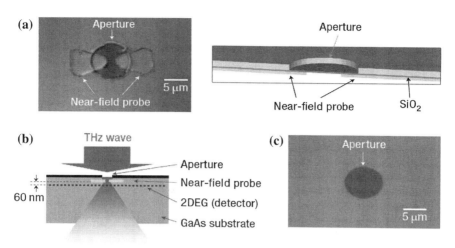

Fig. 1.37 Schematic of the THz-Near filed imaging device. **a** Microscope image of the aperture and planar probes separated by SiO_2 film. **b** Mechanism of THz detection by the imaging system. **c** Optical microscope image of an alone aperture deposited on the surface of a GaAs/AlGaAs heterostructure chip (for comparison) [82]

places the aperture, probe, and detector in close proximity. The 8-μm-diameter aperture and planar probe, each of which is insulated by a 50-nm-thick silicon dioxide (SiO_2) layer, are deposited on the surface of a GaAs/AlGaAs heterostructure chip (Fig. 1.37). A two-dimensional electron gas (2DEG)-located only 60 nm below the chip surface—is used as the terahertz detector. The mechanism of THz detection is based on the resistance change induced by the electron heating effect of the 2DEG where the near-field radiation is sensed before its intensity is reduced by distance. On the other hand, since the detector part comprises a two dimensional layer, it is not affected by the far-field wave of the incident THz radiation. The planar probe changes the distribution profile of the evanescent field and therefore enhances the coupling of the evanescent field to the 2DEG detector which results in strongly enhanced terahertz transmission through the small aperture and detection sensitivity [93]. Numerical calculations of THz electric field distributions have presented in Fig. 1.38 for two situations including the aperture alone and the aperture plus the probe conditions. In the presence of an alone aperture, the electric field is localized close to the aperture while, when the probe is present just behind the aperture, the electric field extends into the interior region of the substrate indicating that the presence of the planar probe changes the distribution profile of the evanescent field, locally enhancing it and thus improving the signal at the 2DEG detector.

The measurement results of the aperture alone and aperture plus probe samples are shown in Fig. 1.38 in order to compare the detection sensitivity of two devices. In the presence of the probe plates, one can see a clear square-wave profile, whereas in the case of an alone aperture, no signal is observed.

Since the CNT detector exhibits much higher sensitivity and has a much smaller sensing area compared with the 2DEG detector, (approximately 200 nm compared

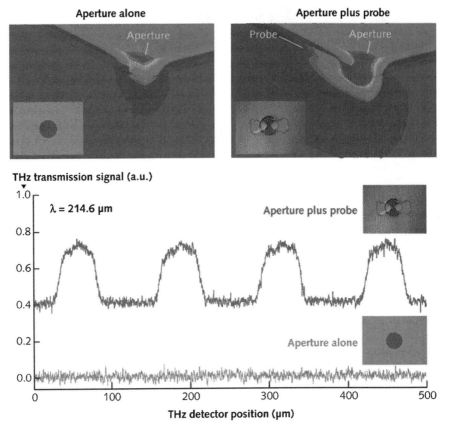

Fig. 1.38 Calculations of THz electric field distributions near the aperture (*upper figure*) and THz transmission signal as a function of the position of the near-field THz detector (*lower figure*) in the presence and without the probe plate. The sample is made up of a terahertz-transparent substrate, the surface of which is covered at regular intervals by terahertz opaque gold films. The widths of terahertz-opaque and terahertz-transparent regions across the scan direction are 80 and 50 μm, respectively and the terahertz transmission profile is measured by scanning the device across a sample [94, 95]

to 8 μm for the 2DEG detector), replacing the 2DEG detector with modified CNT-QD terahertz detector into an on-chip structure for near-field THz detection, would yield high detection sensitivity and high spatial resolution simultaneously.

1.4.1.16 Wide Bandwidth Hot Electron Bolometer Heterodyne Detector Based on Single-Walled Carbon Nanotubes

Hot electron bolometer (HEB) detectors have been developed for low noise terahertz systems intended for space science applications during the last decade [96, 97]. These two-terminal devices employ a heat-able bolometer medium between ohmic contacts where the electron gas is heated to a temperature (T_e) above the lattice temperature, T_0, in response to electromagnetic radiation absorbed in the bolometer. The main advantage of a hot electron bolometer, compared with a standard bolometer, which relies on heating of the lattice, is that the specific heat of the electrons is much smaller than that of the lattice. The bolometer resistance should be a function of the electron temperature to change its resistance as the bolometer is heated. This change in the resistance can be detected by DC biasing the device. Figure 1.39 presents a schematic of a generic electron bolometer device along with the operation principles.

The maximum variation rate of the resistance of a hot electron bolometer is determined by the response time of the electron gas. If the radiative input is turned off, the electron temperature will relax back to the lattice temperature because of heat conduction to the thermal bath, with a time-constant, τ_{TH}. This time constant is referred as the effective response time. The HEB can also be operated in the heterodyne mode by applying two frequencies as the signal or RF frequency (f_{RF}) and the local oscillator frequency (f_{LO}). By absorption of these frequencies by the bolometer, the device can sense the difference (or intermediate) frequency, $f_{IF} = |f_{RF} - f_{LO}|$, and the resistance of the device will vary at this frequency, giving rise to an IF voltage across the device. The 3 dB IF bandwidth can br defined by:

$$f_{3dB} = \frac{1}{2\pi\tau_{TH}} \quad (1.1)$$

The above bandwidth is limited by optical phonon emission. HEBs can be categorized according to the physical processes through which they dissipate the heat generated by the absorbed power. For example, phonon-cooled HEBs utilize a semiconducting 2DEG medium similar to the structure presented in Fig. 1.39, and have a bandwidth of ~3–5 GHz. Diffusion-cooled HEBs (DHEB) on the other hand are cooled by diffusion of the heated electrons into the contacts [98]. The bandwidth of this kind of HEB can be approximated by:

$$f_{3dB} \sim \frac{\pi D}{2L^2} \quad (1.2)$$

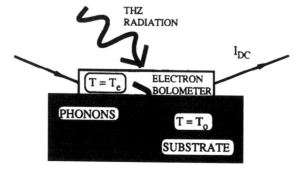

Fig. 1.39 Schematic of a generic bolometer device [96]

Table 1.1 Comparison between different HEB types [98]

#	Type of HEB	Max. IF bandwidth	Bandwidth determined by
1	Insb	1 MHz	Carrier excitation
2	2DEG	3 GHz	Optical phonon emission
3	DHEB Superc.	10 GHz	Electron diffusion
4	PHEB Superc.	5 GHz	Acoustic phonon emission
5	2DEG	20 GHz	Electron diffusion to contacts
6	2DEG	40 GHz	Ballistic electron transport

where D is the diffusivity of the electrons and L the bolometer length. 2DEG HEB cooled by the diffusion of the heated electrons have wider bandwidths. For very high mobility 2DEG material ($>10^6$ cm^2/Vs) it is possible to reach ballistic transport conditions in short bolometers, and the bandwidth then varies as $1/L$, instead of as $1/L^2$. Wide bandwidth of ~ 40 GHz can be obtained using this method [99]. A comparison of different HEB types, their maximum achieved bandwidth and the bandwidth limitation factors are given in Table 1.1.

Temperature-dependence resistance due to ballistic and near-ballistic transport of electron in single-wall carbon nanotubes (SWNTs) is predicted to provide wider bandwidths in the range of hundreds of GHz. The temperature-dependent resistance or bolometric response of a 2 μm metallic SWNT as a function of the power coupled to the tube is depicted in Fig. 1.40. For 2DEG devices with channel lengths of L (between the Ohmic contacts) the measured electron transit velocity can be given by:

$$v_t = 2\pi f_{3dB} L = v_F \qquad (1.3)$$

where $v_F = 2.3 \times 10^7$ cm/s is the Fermi velocity of the 2DEG in the channel. Considering an electron waveguide channel consisted of a metallic SWNT where the Fermi velocity is higher than for 2DEG (8.1×10^7 cm/s) [100] and a device length of 200 nm, a bandwidth of ~ 650 GHz is expected referring to Eq. 1.3. This bandwidth, however, is expected for purely ballistic transport while in the case of quasi-ballistic transport, one can obtain the bandwidth through estimation of the effective transit time, t_{tr}, and using Eq. 1.1.

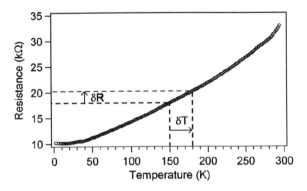

Fig. 1.40 Measured dc resistance as a function of temperature of a 2 μm long quasi-metallic SWNT. THz power coupled to the nanotube leads to a temperature rise δT which produces a measurable resistance change δR [101]

1.4 Terahertz Components

Fig. 1.41 Quasi-optical method for coupling to SWNTs at THz frequencies [102]

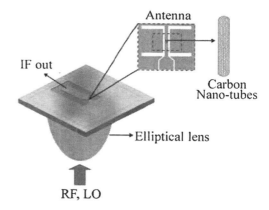

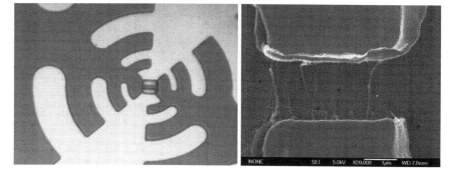

Fig. 1.42 *Left* Optical photograph of an LPA antenna with an etched window for improved CNT placement in the antenna gap. *Right* SEM picture of the gap region of an LPA antenna with several CNT bundles across the gap [103]

In fact, quasi-ballistic SWNTs could provide at least a similar conversion gain as 2DEG ballistic HEBs, with a much wider bandwidth [102]. Terahertz characterization of single SWNTs can be investigated by utilizing a quasi-optical coupling method where an antenna is photolithographically defined in a gold film on a silicon substrate. The substrate is optically coupled via an elliptical or hyperhemispherical silicon lens and the SWNT will be placed and contacted across the gap of the integrated antenna. This arrangement is illustrated in Fig. 1.41. The antenna has an impedance of about 100 Ω while the minimum resistance of a SWNT is about 6.5 KΩ.

Impedance matching between the antenna and the SWNT at the IF frequency (for achieving wider bandwidth) can be realized by capacitive coupling. Place a number of SWNTs in parallel across the antenna gap is another method to achieve matched impedance. To improve the probability of placing the tubes in the antenna gap one may apply photo resist and etch a small window that covered the antenna gap [103]. Figure 1.42 depicts an optical photograph of a log-periodic antenna (LPA) with an etched window. After dissolving the photo resist, the CNTs remain in the gap as presented in the right image of Fig. 1.42.

Due to variety of different types of SWNTs and parameter ranges such as bias voltage, tube length and temperature, the SWNT-based HEBs make it feasible to design imaging systems with arrays consisting of many HEB detectors/IF amplifiers up to 1 THz.

1.4.2 THz Sources

Realization of THz sources using the various techniques such as electronic or photonic techniques faces several serious limitations. Semiconductor based electronic sources, i.e., oscillators and amplifiers are limited by transit times that cause high-frequency roll-off. The problems with tube sources are bulkiness, losses, and high fields as well as high current densities supplies [3]. The energy levels at which photonics sources (solid-state lasers) operates are approximately the same as with lattice phonons energy makes necessary cryogenic cooling. A technique that improved generating terahertz power was frequency conversion. This conversion is either up from millimeter wavelengths, or down from the optical or IR. For example, direct IR laser-to-THz pumping milliwatt source and reactive multiplication through the GaAs Schottky diode milliwatt source operate based on frequency down and up conversion respectively. Several techniques are introduced in microwatt and nanowatt power levels of terahertz sources. Some of these techniques are: optical mixing in nonlinear crystals [104], photomixing (optical difference frequency mixing in a photoconductor coupled to an RF radiator) [105, 106], picoseconds laser pulsing [107], laser sideband generation [108], intersubband and quantum cascade lasing [109], direct semiconductor oscillation with resonant tunneling diodes (RTDs) [110, 111], direct lasing of gases and Josephson junction oscillations [112]. The following compactly discusses up-converters, Tubes, Lasers, and optical down-converters technologies.

Up-converters [3]: Nonlinear reactive multiplication of lower frequency oscillators is the most common technique for producing small amounts of power at frequencies above 500 GHz. The primitive millimeter-wave sources were solid-state (Gunn and IMPATT diodes) and tube sources (carcinotrons, klystrons, and backward-wave oscillators). The most reasonable approach in the source area today is to multiply up from microwave frequencies (20–40 GHz). Although higher frequency (up to 200 GHz) Gunn, IMPATT, and TUNNETT devices are in development but they are not available commercially. Fortunately, advances in III-V monolithic microwave integrated circuit (MMIC) technology extended baseband frequency up to 200 GHz [113, 114]. Mapping the microwave-band to terahertz-band needs a higher order multiplier. Since higher order multipliers ($>\times 4$) have poor conversion efficiencies compared to doublers and triplers, the most efficient terahertz sources are composed of series chains of these lower order multipliers [115]. Multiplied sources most commonly use planar GaAs Schottky barrier diodes mounted in single-mode waveguide. Multiplier chains have reached 1200 GHz with 75 µW at room temperature and over 250 µW when operated cold

1.4 Terahertz Components

(120 K) [116]. Signals up to 2.7 THz have been obtained with this technique [117]. This improvement continues in later efforts.

Tubes, Lasers, and Optical Down-converters [3]: Terahertz tube sources based on emission from bunched electrons spiraling about in strong magnetic fields offer the most power (mW levels) and frequency tuning range at submillimeter wavelengths (extend to 1200 GHz). The main problem with the commercial type of these sources is usually multimoding when coupling these sources to actual antennas. Tube devices are still used for terahertz operations taking the advantages of monolithic fabrication techniques and high-density cathode development [118–120].

IR-pumped gas lasers based on grating tuned CO pump lasers (20–100 W) injected into low-pressure flowing-gas cavities can produce the terahertz signals. Depending on the chosen emission line, the power level may be in 1–20 mW range. A combination of lasers and harmonic generators can be used to make more tunable sideband sources. Lately direct semiconductor terahertz laser sources based on intersubband transitions and tailored quantum cascade laser devices are in the center of attention in the tournament with other terahertz laser sources [121]. Despite some difficulties (such as requiring cooling) these devices are achieving significant CW power sources thanks to ever-expanding list of compound semiconductor materials.

One of the most successful techniques for generating terahertz is energy-down-conversion from the optical regime. Two techniques are used to produce narrow- and broad-band energy. The technique to produce narrow-band power is photomixing. In this technique, offset-frequency-locked CW lasers are focused onto a small area of a photoconductor to generate carriers between closely spaced electrodes printed on the semiconductor. The photocurrent produced by photocarriers which is modulated at the laser difference frequency is coupled to an RF circuit or antenna to radiate the terahertz energy. For a single device photomixer, optical to terahertz conversion efficiencies are low, however, cooling, arraying, and new materials can help these devices improvement.

To produce broad-band terahertz energy an optical technique based on using a femtosecond pulse optical laser is used [122]. Through this technique carriers are generated by lasers between closely spaced electrodes on a photoconductor and then accelerated in an applied field. RF power may be radiated by coupling the resulting current surge to an RF antenna. The radiated energy has frequency components that reflect the pulse duration, i.e., terahertz rates.

In this optical technique, on the other hand photoconductor can be replaced by a crystal with a large second-order susceptibility like zinc telluride. The nonlinear susceptibility leads to mixing. The occurred mixing produces a time-varying polarization with a frequency–response representative of the pulse length, i.e., terahertz oscillation. Average output power levels of these broadband terahertz sources over the entire spectrum falls within the range of nanowatts to microwatts.

For high levels of terahertz output power, another pulsed laser technique was developed that uses a Q-switched Nd:YAG laser to illuminate a large $LiNbO_3$ sample causing optical parametric oscillation [123]. Two photons are created through the parametric process: one near-IR photon (close to laser pump wavelength) and one terahertz difference photon. The crated terahertz photons are extracted by a prism coupler.

In the following, certain techniques, technologies and special cases will be discussed in detail.

1.4.2.1 The Development of Terahertz Sources: The Search Continues for Efficient Terahertz Sources [124]

Traditional Terahertz Sources: Terahertz emitters can be realized using microwave technology based on Gunn, Impatt, or resonant tunneling diodes. However, the main challenge in dealing with the mentioned sources relies in the operation frequency of these devices which is not high enough for many applications. Therefore, special mixers are required to multiply their frequencies.

Synchrotron sources, free-electron lasers, the Smith-Purcell emitter, and backward-wave oscillators are a less-compact class of terahertz emitters based on an electron beam. In a synchrotron and free-electron laser, oscillatory movement of electrons in an alternating magnetic field concludes to terahertz radiation. Both of the sources have high frequency tenability and output power however they both suffer from very large dimensions.

The Smith-Purcell emitter is nearly a compact terahertz source where the propagation of electron beam from an electron microscope along the surface of a metallic grating results in a magnetic field effect and a similar phenomenon like that of the free-electron sources appears. Backward-wave oscillators or carcinotrons are another type of terahertz sources that electrons are grouped in periodic branches by flying over a comb-like structure and emit terahertz radiations. Although these sources are tunable monochromatic devices they suffer from relatively large dimensions too. Generation of terahertz waves by using the transitions between different rotational states of gases in molecular gas lasers is another method that uses traditional laser sources. Due to existence of transition-rich gases such as methanol, methyl fluoride, formic acid, etc. a wide frequency range from less that 300 GHz to more than 10 THz can be covered by these sources. The output power of these compact lasers can reach to several milliwats however they are relatively expensive devices for most applications. Semiconductor lasers have solved both of the compactness and price challenges of the molecular gas lasers but cryogenic cooling is their main operation problem. P-germanium lasers in which hole transitions from the light-hole band to heavy-hole band is used to generate THz radiation are an example of semiconductor lasers. The use of these sources is limited due to their low-temperature and magnetic-field-dependant operation.

The advent of femtosecond lasers in the early 1990s was a great accomplishment in detection and generation of terahertz pulses. The basis of these lasers relies on photoconductive dipole antennas that are gated by the femtosecond pulses. The antennas consist of a metallic antenna structure deposited on a semiconductor substrate. An externally applied bias forms a current that can only flow under the action of a laser pulse that optically excites the semiconducting antenna substrate in the gap. Finally, a short current pulse with a subpicosecond rise time is obtained due to acceleration of generated electron–hole pairs in the bias field which is source of short (broadband) terahertz pulses.

Quantum cascade lasers (QCL) are another promising technology to realize room-temperature, monolithic and compact terahertz sources especially in the range of 1–5 THz which rely on the emission from intersubband transitions in a quantum well [125]. Consisting of two separate active and relaxation regions that are grown successively, electrons emit terahertz photons in each transition inside the active region. Early QCLs required cryogenic cooling and emitted in midinfrared wavelength and today CW mid-IR QCLs have found industrial applications [126]. However, recent progresses have lead to advent of QCLs working at 4.4 THz [127] and 3.2 THz [128].

Terahertz-wave parametric generators are another compact but clearly room-temperature class of terahertz sources which work base on the parametric conversion of nanosecond laser pulses from Q-switched, diode-pumped Nd:YAG lasers [129]. These sources exhibit narrowband, tunable and so powerful radiation that can be detected at room temperature with a pyroelectric detector.

Solid State Terahertz Sources: For THz frequency range, achieving to room temperature solid-state sources is a great deal. In lower frequencies (microwave), fabrication of electronic devices to operate at frequencies above a few hundred GHz has been difficult [130] (although the output of such sources can be harmonically multiplied to the THz range [131]). This is consequence of the inherent need for very short carrier transit times in the active regions. In higher frequencies (optical), it is even difficult to generalize the concept of interband diode lasers at visible and near-infrared frequencies into mid-infrared because appropriate semiconductors are not available.

Recent improvements in the field of the quantum cascade (QC) lasers in which inter-subband or inter-miniband transitions in layered semiconductor heterostructures are used to generate light in the mid-infrared region have been promising [132]. These sources can cover a wide range of wavelengths, from the far- to the mid-infrared and room-temperature operation at up to the longest wavelength of 24 µm (12.5 THz) [133] by tailoring the width of quantum wells in the active region. However, extending this concept to frequencies below 10 THz is a great challenge since free carrier absorption in semiconductors increase with the wavelength and also the small energy related to these terahertz radiations (10–20 meV) of the optical transition requires very selective injection. On the other hand, there are challenges in designing waveguides for THz radiations.

Because of the mentioned difficulties in fabricating solid-state THz sources, generation of THz radiation through all-optical techniques such as visible/near infrared femtosecond pulsed lasers has attract great concern. These coherent THz systems have been used in demonstration of THz radiation for imaging and spectroscopy applications. So, it would be interesting to review the mechanisms of converting ultra-short (~ 100 fs) near-infrared pulses into THz pulses.

Coherent terahertz imaging and spectroscopy systems: THz pulses are generated using a semiconductor THz generator which converts the incident ultra-short ($\sim$ 10–200 fs) visible/near-infrared pulses are into THz radiation. The THz pulses are either emitted co-linearly with the reflected incident pulses, or propagate through the semiconductor generator and emerge from the far side. The emitted THz radiation is collected and collimated by an off-axis parabolic mirror, and then focused by a second off-axis parabolic mirror onto the sample, which can be stepped across the beam to build up a two-dimensional image. A schematic setup of a THz time-domain spectroscopy system is presented in Fig. 1.43 [134].

Several methods have proposed to coherent detection of transmitted or reflected THz radiation [135]. The two-dimensional THz imaging setup shown in Fig. 1.43 is a technique based on the ultrafast Pockels effect [136]. In this technique, the THz beam is collected and focused onto an electro-optical sampling (EOS) detection crystal such as ZnTe and causes an instantaneous birefringence in this medium which is probed with a second visible/near-infrared beam. The birefringence modulates the ellipticity of the probe beam. A combination of a quarter wave plate ($\lambda/4$), a Wollaston polarization (WP) splitting prism, and two balanced photodiodes is then utilized to measure the ellipticity of probe beam. The photodiode signal can be measured by placing an acousto-optic modulator in the pump beam and using lock-in methods. Also, the electric field of the THz pulse in the time domain can be obtained through measuring the photodiode signal as a function of the time-delay between the arrival of the THz and probe pulses at the EOS crystal. This field is presented in Fig. 1.44. Information about the frequency spectrum of THz radiation can be obtained through Fourier transform as depicted in Fig. 1.44.

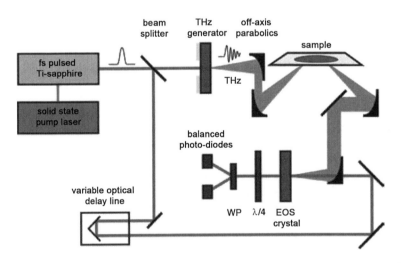

Fig. 1.43 Schematic setup of a two-dimensional THz time-domain spectroscopy system [124]

1.4 Terahertz Components

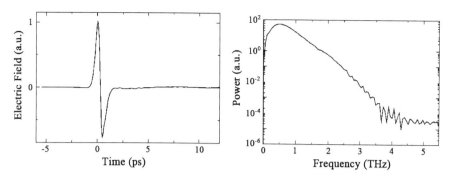

Fig. 1.44 *Left* Electric field of the THz pulse obtained through spectroscopy system in the time domain. *Right* Frequency spectrum of the THz pulse achieved from Fourier transformation [124]

THz Semiconductor Sources: Semiconductor surfaces have been widely used in conjunction with femtosecond visible/near infrared lasers as THz emitters, exploiting bulk electro-optic rectification (difference frequency mixing) and ultrafast charge transport techniques.

- Bulk electro-optic rectification

It is possible to generate the THz radiation through exploiting the second order nonlinear susceptibility ($\chi^{(2)}$) of a semiconductor by applying a large peak electric field corresponding to a visible/near-infrared pulse. In this case, a time-dependent polarization given by $P(\omega_{THz}) = \varepsilon_0 \chi^{(2)} E(\omega_{vis1}) E(\omega_{vis2})$ proportional to the intensity of the incident pulse is induced which is in the THz frequency range ($\omega_{THz} = |\omega_{vis1} - \omega_{vis2}|$). Due to short temporal duration of the incident pulse, a broad bandwidth (>10 THz) may obtained. Terahertz radiation have been generated in a number of organic and inorganic semiconductors such as ionic salt 4-*N*, *N*-4-dimethylamino-4'-*N*'-methylstilbazolium tosylate (DAST) [137] and GaAs, GaSe and ZnTe [138–141] using this technique. The efficiency of this process is determined by the magnitude of the induced polarization or in the other words by the magnitudes of the second order susceptibility [138]. Also, the phase matching condition between the induced THz field and the optical fields determines the conversion efficiency. This condition denoted by $k_{vis2} = k_{vis1} \pm k_{THz}$ can result in a major limitation for this technique.

- Ultra-fast charge transport

Photo-excited electron–hole pairs in semiconductor structures can be used as another THz source. Electron–hole pairs are generated close to the surface of the generation crystal by an ultrafast visible/near-infrared pulse with photon energy greater than the semiconductor band-gap and then are accelerated in an electric field and the subsequent dipole changes result in generation of THz pulses.

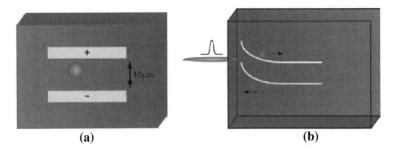

Fig. 1.45 Schematic diagrams showing two THz emitters in which photoexcited carriers generated by a focused visible/near-infrared laser pulse are accelerated and radiate in an electric field. In **a** the field is provided by a lateral antenna structure, in **b** the intrinsic semiconductor surface depletion/accumulation field is exploited [124]

Two methods have been proposed to realize the surface fields; lateral antenna structure and exploiting the intrinsic semiconductor surface depletion or accumulation. In the first method, a lateral antenna comprising two electrodes is deposited on a semiconductor surface. A large electric field is applied between the electrodes which accelerate the photo-excited carriers generated by the incident laser pulse focused between the electrodes [142, 143]. This method is schematically shown in Fig. 1.45a. In the second method depicted in Fig. 1.45b ultrafast carrier transport is driven by both the intrinsic electric field perpendicular to the semiconductor surface [144, 145] and by differences in the electron and hole mobilities (photo-Dember effect) [146, 147]. This method has obtained much attention in recent years.

Several parameters such as temperature, energy, width and flux of the incident laser pulse and properties of semiconductors determine the emitted power and bandwidth of the generation. An external magnetic field can be used to enhance the emitted THz power by over an order of magnitude [148–152].

Surface Field Generation: Generation of THz radiation through ultrafast charge transport perpendicular to a semiconductor surface have been utilized in emerging applications of coherent THz spectroscopy [153] and imaging in recent years. This method also provides a powerful, non-contact method to study physical phenomena like hot carrier dynamics [154] and collective processes with sub-picosecond time resolution.

THz emissions of three different semiconductors (GaAs, InAs and InSb) excited by 2 nJ pulses of 1.6 eV photons from a mode-locked laser are presented in Fig. 1.46 as a function of magnetic field [149, 150, 152]. The emitted THz radiation was collected parallel to the applied magnetic field and perpendicular to the incident beam. The experimental geometry of this experiment is illustrated in Fig. 1.47. It can be concluded that the dependence of the *TM* and *TE*-polarized THz fields on the applied magnetic field, and the maximum emitted THz power, differs significantly between semiconductor surfaces.

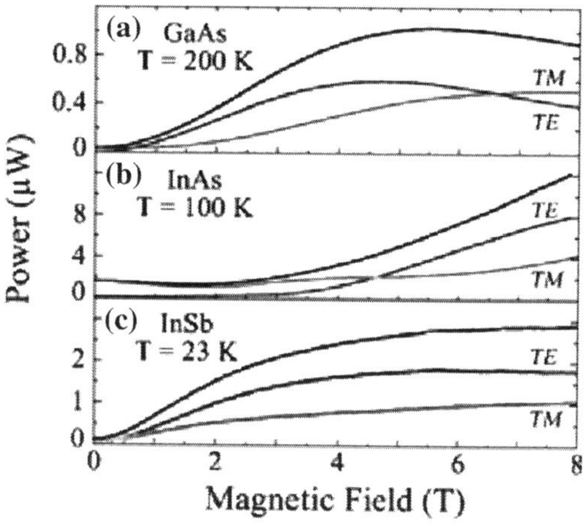

1.4 Terahertz Components

Fig. 1.46 Emitted THz power versus applied magnetic field for bulk GaAs, InAs, and InSb including TE/TM components and the total radiated power [124]

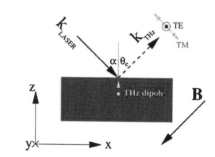

Fig. 1.47 Experimental geometry. An fs laser pulse is incident on the semiconductor surface at angle α, leading to a THz dipole. The emitted THz radiation is measured at angle θe. The direction of the applied magnetic field is shown by the vector **B**. Figure 1.46 data has $\alpha = \theta e = 45$ [124]

1.4.2.2 Terahertz Sources Based on Schottky Diode Frequency Multipliers [155, 156]

Considerable improvements have been obtained during the last years in developing diode frequency multipliers with acceptable output power in terahertz region. These elements generate harmonics of an input signal through utilizing the reactive and resistive nonlinearity of the diode. GaAs Schottky diodes are examples of the dominant technology for terahertz frequency multipliers which are developed on membranes with a thickness of a few micrometers. Achieving to electronically tunable sources with sufficient output power have been greatest challenges in the developing these frequency multipliers. Waveguide biasness balanced frequency tripler multipliers have been proposed to cover the 1.6–1.7 THz and 1.7–1.9 THz bands by using two Schottky diodes on GaAs membrane as displayed in Fig. 1.48 [157, 158].

An E-plane probe is located in the input waveguide on multiplier circuit which couples the signal at the input frequency to a suspended microstrip. This line matches the diodes impedance at the input and output frequencies and prevents the leaking of third harmonic to input waveguide. The diode-generated third harmonic

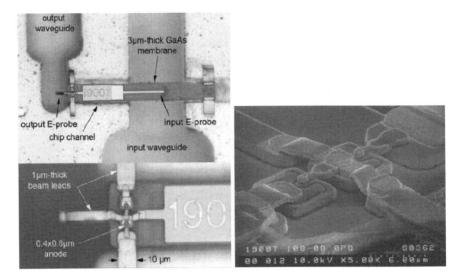

Fig. 1.48 Frequency tripler for the 1.6–1.7 THz band and 1.7–1.9 THz local oscillator chains of the heterodyne instrument of the Herschel Space Observatory. *Left* Multiplier chip placed inside the waveguide block. *Right* Close-ups image of the chip [155]

is then coupled to the output waveguide through another E-plane probe. The dimensions of the output waveguide and the balanced geometry of the chips cut off the second harmonic and suppress the power of the fourth harmonic of the input, respectively.

Simulations with realistic assumptions for diode parameters describe that these multipliers have the capability of operation above 1 THz [159].

The measured efficiency of the frequency multipliers can be fit by the following equation [156]:

$$\eta(f) = \eta_0 \cdot e^{f/f_0} \tag{1.4}$$

where η_0 and f_0 are specific constants for doublers and triplers that depend on the operation temperature and the bandwidth. The efficiency of N cascaded doublers or triplers can also defined by:

$$\eta(f) = \eta_0^N \cdot \exp\left[\frac{f}{f_0} \cdot \sum_{i=0}^{N-1} 2^{-i}\right] \tag{1.5}$$

where the summation is over the cascaded multipliers. It is clear that the constants η_0 and f_0 have been reduced effectively compared with a single multiplier but the total efficiency is exponentially decaying similar to Eq. 1.4. The two 1.6–1.7 THz and 1.7–1.9 THz chains of the frequency multiplier presented in figure 1 have an output power of 21 and 4 µW at room temperature corresponding to conversion efficiencies of 1.5% at 1647 GHz and 0.4%

1.4 Terahertz Components

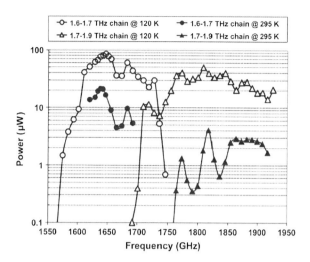

Fig. 1.49 Output power of the frequency multiplier presented in figure 1 for the 1.6–1.7 THz and 1.7–1.9 THz chains at room temperature and at cryogenic temperature (120°K) [155]

at 1818 GHz, respectively. Also, the output powers of the two chains are about 100 μW and 50 μW for 1.6–1.7 THz and 1.7–1.9 THz chains at cryogenic temperature (120°K), respectively. The output powers versus the frequency variation have depicted in Fig. 1.49 for two temperatures. It is obvious that at lower temperatures the output power and also the operation bandwidth increase considerably. This is because of enhanced mobility of GaAs, decreased ohmic waveguide and the on-chip matching circuits' losses due to decreased phonon scattering, and finally increasing the efficiency of the last stage upon increasing the drive power.

Efficiency dropping power and short device lifetime are the main limitations in increasing the output power of the frequency multipliers. The first problem occurs due to saturation effect while the second challenge relates to thermal or reverse breakdown effects [160]. Increasing the number of anodes per chip is one of the considered methods to increase the output power. Frequency triplers at 300 and 600 GHz with respective six and four Schottky anodes per chip are samples demonstrated according to this method [161, 162]. Increasing the number of anodes deals with several challenges that should be addressed. For example, there has to be trade-offs between an optimum coupling to the anodes, an optimum matching of each anode at the idler frequencies, and optimum matching at the output frequency.

1.4.2.3 Free Electron Based Terahertz Sources [163]

Klystrons, Backward Wave Oscillators (BWO), Travelling Wave Tubes (TWT), and Gyrotrons are examples of free electron based sources that have been investigated to reach to high frequency part of the microwave region. However, physical scaling problems, metallic wall losses and the need for high magnetic and electric

Fig. 1.50 Output power and spectral distribution of BWO devices [163, *Courtesy of* M. Dressel, University of Stuttgart]

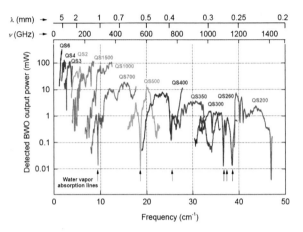

fields are the challenging points that these devices have been faced. Between these free electron terahertz sources only Gyrotrons have attracted attention due to the capability of generating high average power. TWTs on the other hand have been considered as candidate sources of radiation above 100 GHz. The BWO is the other free electron slow wave device that can operate in the terahertz region with an accelerating potential of about 1–10 kV and axial magnetic field of about 1 T and moderate output powers (1–100 mW) [164]. Changing the accelerating potential can tune the BWO over tens of GHz. In fact, in this device electrons spiralize through a corrugated structure in the presence of a magnetic field. Therefore, they interact with the first spatial harmonic of the backward wave. In this region of the dispersion relation, the phase velocity of the wave is positive and the group velocity is negative. Different BWOs can be implemented in an integrated system to cover a wide frequency range (from 30 GHz to 1.2 THz). Figure 1.50 depicts the output power and frequency distribution of BWO devices.

1.4.2.4 Compact Tunable Terahertz Sources: Very Short Wavelengths Vacuum Electronic Devices (Folded Waveguide Travelling Wave Tube) [165]

Although semiconductor based terahertz sources, consisting of electronic devices such as the Gunn diode, or of optical lasers such as the quantum cascade laser, have demonstrated a new vision for different applications but a semiconductor based source with simultaneous high output power, high efficiency and high bandwidth has not developed yet. Vacuum electronic devices such as the gyrotron or the free electron laser, on the other hand, suffer from several drawbacks. They are usually large and expensive, and require very high voltage for their operation.

Combination of the properties of vacuum tubes with solid-state micro-fabrication methods can be an alternative for the next generation of terahertz wave devices [166]. The high efficiency infinite electron mobility are of main advantages of vacuum devices compared with solid-state components where carrier

1.4 Terahertz Components

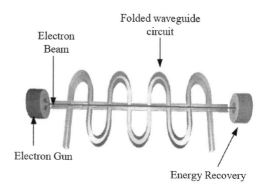

Fig. 1.51 FWTWT THz system [165]

mobility is a fundamental limitation and leads to device heating problems and therefore reduced efficiency.

Beyond the problems associated with solid-state components, solid-state fabrication methods are economic for mass production and lead to micron size accuracy. In recent years, realization of vacuum electronic devices through these methods has become possible. Moreover, silicon electron guns which operate at much lower temperatures and produce higher beam current densities than are achievable with conventional thermionic emitters have been manufactured recently. Terahertz range reflex klystrons [167] and millimeter range klystrons [168] are examples of the devices that the solid-state fabrication methods have applied.

However, the travelling wave tube in the microwave range offers greater bandwidth and flexibility than the klystron and also realization of folded waveguide travelling wave tube (FWTWT) is possible through micro-fabrication techniques [169]. Figure 1.51 illustrates a schematic sketch of a FWTWT. In an amplifier configuration, a 2.5 cm long device fabricated in silicon can achieve 10 dB of gain at 500 GHz, with a saturated output of about 100 mW. With the application of some positive feedback this amplifier can easily be converted into an oscillator that can possibly fill the Terahertz Gap.

1.4.2.5 Photomixing Tunable Terahertz Sources [170]

Photoconductive mixing or photomixing is a useful method for coherent generation in the THz region. This technique requires pumping an ultrafast photoconductive diode with two frequency-offset lasers. In the other words, photomixing is generation of continuous wave terahertz radiations by means of two mixed continues wave lasers with identical polarisations which are focused onto a photomixer device. Photomixing is technologically a considerable method since there are a few tunable sources capable of generating terahertz radiations. In this technique, spatial overlapping of two lasers with the frequencies ω_1 and ω_2 is used to generate a terahertz beam. After that, the co-linear lasers illuminate an ultrafast semiconductor (e.g. GaAs). Low temperature-grown of semiconductor material such as GaAs or ErAs:GaAs, have lead to realization of ultrafast photoconductive diodes with very short carrier lifetimes (~ 0.25 ps) [171, 172]. Low-temperature growth of the mentioned materials also provides suitable photocarrier mobility which can reach to 100 cm^2/V s for electrons and holes compared with other ultrafast materials such as defective silicon.

Due to photon absorption and short carrier lifetime, the conductivity of the material can be modulated at the tuned terahertz frequency $\omega_{THz} = \omega_1 - \omega_2$. Then, variation of the conductivity can be converted to electrical current and finally radiation through an antenna pair by applying an electric field. The tenability of the photomixer can be enhanced further by coupling the photoconductive diode to a multi-octave antenna like logarithmic spiral which results in broadband photomixers [173]. This high frequency tenability along with the room-temperature operation capability is the main advantageous of photomixing compared with other solid state terahertz sources. A micrograph of a fully fabricated photomixer with an enlargement of the interdigitated electrodes is presented in Fig. 1.52.

This structure consists of three epilayers where the first layer is an AlAs/AlGaAs distributed Bragg reflector (DBR), the second layer is an AlAs heat spreader and the third layer is an ErAs:GaAs layer. The ErAs creates nanoparticles inside the GaAs layer which act as fast non-radiative recombination centers without significant degradation of the mobility. Also, the DBR layer reflects the wavelength 760 nm but passes terahertz radiations.

An experimental setup for the THz photomixer is presented in Fig. 1.53. A dc power supply is used for biasing the photomixer and a tunable master oscillator power amplifier laser (tunable within a fraction of GHz) along with a fixed laser are used to generate the THz beatnote. The output power of the lasers is split with a beam splitter. A part of the split beam is coupled to an optical spectrum analyzer (OSA) which allows the continuous monitor of both laser wavelength and accordingly the THz difference frequency while the other part is focused on the

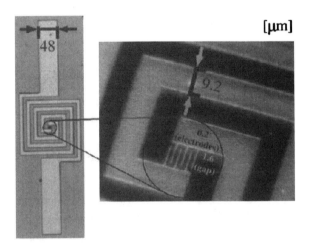

Fig. 1.52 Image of a spiral antenna (*left*) and interdigitated fingers (*right*) [170]

1.4 Terahertz Components

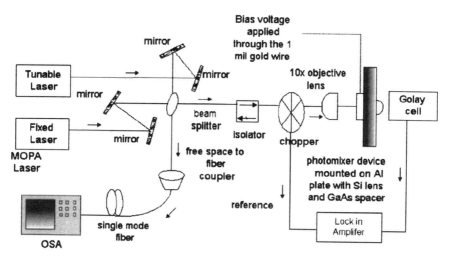

Fig. 1.53 Experimental setup used to characterize ErAs:GaAs photomixers [170]

photomixer. A composite dipole-spiral antenna radiates the photomixer generated ac current to free space through hyperhemispherical Si lens and the radiation power is sensed via a Golay cell.

Considering the operation specifications such as room-temperature operation, tenability and coherence of the photomixer, this THz source is expected to have various applications such as THz spectroscopy, imaging, bioparticle detection, etc.

1.4.2.6 Terahertz Magnetic Response from Artificial Materials

Artificial materials with engineered electric and magnetic response have been topic of interesting researches in recent years and artificial magnetic structures have been fabricated that exhibit magnetic response at terahertz frequencies. These artificial materials, also known as metamaterials or left-handed materials, exhibit negative permittivity and permeability which result in specific optical properties in these materials. The excitement about electromagnetic metamaterials stems from the ability of these materials to exhibit an electromagnetic response which is not observable in natural materials such as negative refractive index [174] and artificial magnetism [175]. However, it is necessary to design artificial materials such that they exhibit an effective material response to electric and magnetic fields [176]. A pair of concentric split rings, often referred to as a split ring resonator (SRR) is the most common element to obtain a magnetic response also this structure displays an electric response under normal incidence radiation with the magnetic field lying completely in the SRR plane and electric field perpendicular to the SRR gap. This allows the electric field to drive the inductive-capacitive (LC) resonance. An incident wave with magnetic field vector parallel to the axis of an SRR will result in an electromagnetic force in the rings which leads to current flow

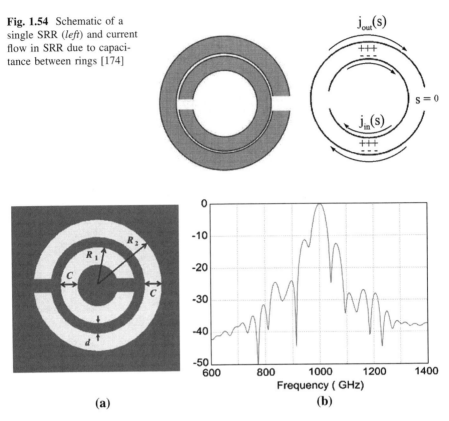

Fig. 1.54 Schematic of a single SRR (*left*) and current flow in SRR due to capacitance between rings [174]

Fig. 1.55 a Split ring resonator geometry, and **b** frequency response of SRR structure centered at 1 THz [177]

in each ring due to the large amount of capacitance between the inner and outer rings as illustrated in Fig. 1.54.

The SRR geometry has unique specifications such as high Q (e.g. thousands), wide fee spectral range (e.g. 6 THz), and straightforward fabrication and depending on the structure dimensions, frequency response at different THz range has been obtained. For structures on the order of micron to submicron and fabricated through electron beam lithography, frequencies in the neighborhood of 100 THz are reported. Also, for larger geometries (fabricated by standard optical lithography), frequencies in the range of 10s of THz are generated. Simulation result for a SRR structure with geometry parameters given by $R_1 = 60$ μm, $R_2 = 110$ μm, $C = 30$ μm, and $d = 20$ μm, is presented in Fig. 1.55 where the structure has designed to achieve 1 THz operation.

In principle, the SRR can also be used as an electrically resonant particle as it exhibits a strong resonant permittivity at the same frequency as the magnetic resonance. However, the electric and magnetic resonant responses are coupled,

1.4 Terahertz Components

resulting in rather complicated bianisotropic electromagnetic behavior [178]. Designing symmetric structures usually referred as electronic split ring resonator (eSRR) which is composed of two single rectangular rings joined at the split gap, can lead to elimination of magneto-optical coupling effects related to bianisotropy and suppress the magnetic response and hence, yields electrically resonant structures. Simulation and experimental results for a number of electric metamaterials structures including simulated norm of the electric field at resonance (column (a)), the surface current density (column (b)), and the experimentally measured field transmission, T(ω), are presented in Fig. 1.56. The absence of a magnetic response can be inferred from the surface current densities since the magnetic fields created by circulating surface currents cancel due to clockwise and counterclockwise components in adjacent regions of the structures.

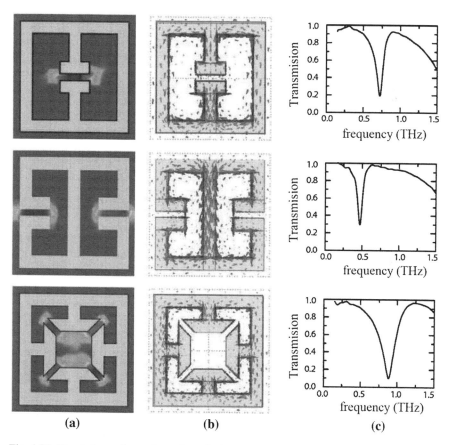

Fig. 1.56 Simulation and experimental results for electric metamaterial structures: **a** the norm of the electric field at resonance, **b** the surface current density, and **c** experimentally measured transmission [179]

Considering the lack of intrinsic response from natural materials at terahertz frequencies (THz gap), the electric metamaterials structures may play an important role in filing the THz gap.

The initial demonstration of these structures at THz frequencies highlights their usefulness and versatility.

1.4.2.7 Continuous-Wave THz Radiation Generation Through Nonlinear Processes

During the past years several methods have been proposed for generation of terahertz waves mainly based on pumping with Q-switched or femtosecond laser pulses resulting in pulsed THz waves which have limited applications. On the other hand, despite of potential for numerous applications for CW THz waves, a few methods have proposed for development of these sources [180–182]. CW THz waves can be used for carriers and modulators in telecommunications. Development of high power, tunable CW THz sources operating at room temperature are necessary to enable practical applications, especially in spectroscopy and imaging technologies. Also, CW THz waves pumped using semiconductor lasers have very narrow linewidth (of megahertz order) and can obtain high-resolution THz spectra rapidly. Nonlinear-optical methods have been considered for generation of monochromatic terahertz radiation in recent years [183, 184]. Difference frequency generation (DFG) and optical parametric oscillation (OPO) are of the attended methods for this purpose. The output power of the generated THz waves in DFG in a non-collinear configuration is related to the input laser powers as:

$$P_{\text{THz}} = A\left(\frac{P_1 P_2}{S}\right) L_{\text{eff}}^2 \tag{1.6}$$

where P_1 and P_2 are the effective powers of the pump and signal beams, L_{eff} is the effective interaction length of near-IR beams in non-collinear phase-matched mixing, S is the cross-sectional area of the pump and signal beams and A is a material constant for generating THz waves from a specific crystal such as GaP. It is clear that the THz power can be enhanced by increasing the pump and signal powers, increasing the effective interaction length, L_{eff}, of near-IR beams in the GaP crystal and decreasing the cross-sectional area of the near-IR beams, S. The experimental setup for generation of CW THz wave using DFG in GaP crystal is shown in Fig. 1.57 where a distributed feedback (DFB) laser and an external cavity laser diode (ECLD) with angular frequencies of ω_1 and ω_2 as the pump and signal lasers have utilized. Laser diodes exhibit narrow linewidth (a few megahertz) and have wide frequency tunability with stable output power and are utilized with fiber amplifiers (FAs) to amplify the output power of the lasers to increase P_1 and P_2. The wavelength of DFB laser tunes from 1058 to 1061 nm with a linewidth of 2 MHz by changing the temperature of the diode and the wavelength of

1.4 Terahertz Components

Fig. 1.57 Schematic of the experimental setup used for generation of CW THz waves in GaP crystal [183]

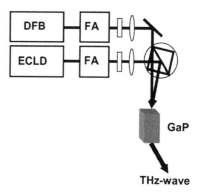

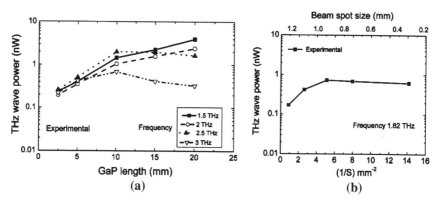

Fig. 1.58 Generated THz wave power as a function of (a) GaP crystal length at 1.5–3 THz, (b) inverse of spatial overlapping of cross-sectional areas of pump and signal lasers at 1.62 THz and beam spot size [183]

the ECLD tunes from 1020 to 1080 nm with a linewidth of 2 MHz and consequently the terahertz wave is generated at $\omega_{THz} = \omega_1 - \omega_2$.

The experimentally obtained output THz power as functions of GaP crystal length, inverse of spatial overlap of the cross-sectional area of input lasers and beam spot size are presented in Fig. 1.58a, b. An output power of 4 nW is achieved with a 20 mm long GaP crystal at a frequency of 1.5 THz. Despite of wide frequency tunability and narrow linewidth of the generated THz wave, the output power is still low and requires further enhancement for certain applications

Although OPOs are more versatile processes compared with the DFG due to their tuning characteristics, they suffer from high power threshold needed for oscillation generating terahertz waves. The threshold power can reach to several hundred of watts due to high absorption of terahertz radiation by vibrational excitations. Cascaded nonlinear processes within an optical cavity are expected to result in intensity enhancement up to the level required for parametric oscillations [185]. Considering the resonance condition, $\omega_p = \omega_s + \omega_i$, and the phase-matching condition for OPO given by:

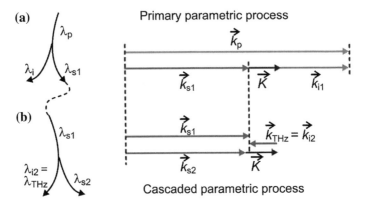

Fig. 1.59 a Primary parametric process and b cascaded parametric process and generation of the terahertz wave [185]

$$\vec{k_P} = \vec{k_S} + \vec{k_i} + \vec{K} \quad (1.7)$$

where ω_P, ω_S and ω_i are the angular frequencies and k_P, k_S and k_i are the wave vectors of pump, signal and idler, one can select K (grating vector of an alternating second-order nonlinearity induced by periodic poling of crystals) such that the quasi-phase matching occur. In the cascaded parametric process the pump wave (k_P) converts to a signal (k_{S1}) and an idler wave (k_{i1}) in a primary process. The signal wave (k_{S1}) resonantly enhances by the cavity and acts as a pump wave for the cascaded process and generates a second signal (k_{S2}) and idler (k_{i2}) waves in the presence of the grating vector, K. The second idler wave is in the terahertz rang, i. e. $k_{i2} = k_{THz}$. These processes are schematically illustrated in Fig. 1.59.

Since the signals have very close frequencies, the cavity is resonant for both of them and according to phase-matching condition satisfied through equation 3, the terahertz wave propagates backward:

$$|\vec{k_{THz}}| = |\vec{k_{S2}}| - |\vec{k_{S1}}| + |\vec{K}| \quad (1.8)$$

The experimental setup for generation of the terahertz radiation consists of a singly-resonant optical parametric oscillator with a bow-tie cavity pumped by a CW laser as depicted in Fig. 1.60. The cavity includes two highly reflecting (>99.9%) concave and plane mirrors and an off-axis parabolic mirror is placed directly after the first concave mirror which deflects out the terahertz wave and transmits pump and signal waves. In order to detect the deflected terahertz wave, the beam is sent to a second off-axis parabolic mirror and is focused onto a Golay cell.

Measurement results describe that three different operation regions can be assigned according to the input pump power. For the pump powers below 2.8 W no oscillations occur while for the power higher than this limit, the primary process starts. The secondary process then starts at 4.7 W. Figure 1.61 shows the spectra of

1.4 Terahertz Components

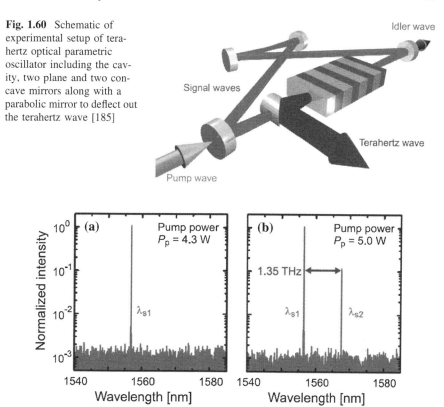

Fig. 1.60 Schematic of experimental setup of terahertz optical parametric oscillator including the cavity, two plane and two concave mirrors along with a parabolic mirror to deflect out the terahertz wave [185]

Fig. 1.61 Spectra of the signal waves for pump powers of (**a**) 4.3 W (the secondary parametric process has not started yet), and (**b**) 5 W (the terahertz wave is generated at 1.35 THz) [185]

the signal waves at two different pump powers. For input pump power of 4.3 W the secondary process has not started yet and only the signal wave of the primary parametric process, λ_{s1}, is present at 1557 nm while for a pump power of 5 W signal waves of the primary and the cascaded parametric processes, λ_{s1} and λ_{s2}, appear. The frequency separation of the two signals denotes the generated terahertz wave at 1.35 THz with an output power higher than 1 μW. The obtained terahertz frequency can be tuned by varying the period length of the poling structure and also by changing the crystal temperature from 1.3 to 1.7 THz.

1.4.2.8 Terahertz Source Using Triply Resonant Nonlinear Photonic Crystal Cavities (PhCNC) [186]

Generation of coherent terahertz radiation through nonlinear processes is one of the promising methods that have some advantages over traditional techniques [187–192]. In traditional methods such as using photoconductive antenna [193] or optical rectification (OR) of amplified femtosecond pulses in ZnTe crystals [194] reported previously for generation of THz radiations, an ultrafast laser with high peak power and macroscopic optics are required which makes these methods not suitable for most of applications. Using quasi-phase-matched LiNbO3 [195] and GaAs [190, 196] crystals in the OR approach to increase the nonlinear interaction between the pump and generated signals, has result in a considerable photon conversion efficiency of 3.3% [190].

Cavity-enhanced nonlinear frequency conversion is another method to generate THz radiation via nonlinear processes that have attended in recent years [197]. Figure 1.62 presents a schematic of a triply resonant structure consisted of two nested photonic crystal cavities that is capable of THz generation based on second order nonlinear susceptibility, χ^2. The first cavity which is demonstrated in a nonlinear medium like III–V semiconductors produces nonlinear polarization at terahertz frequencies from input near-infrared wavelengths according to DFG. NIR pump and idler waves are coupled into the cavity via a waveguide extending from one end of the NIR cavity. This cavity is in the vicinity of a larger single-mode cavity with the resonance in the THz range. Since the wavelengths of the pump and the idler, λ_1 and $\lambda_2(\sim 1.5\,\mu m)$, are completely different from the terahertz radiation wavelength, $\lambda_T(\sim 1.5\,\mu m)$, the two cavities can be considered to be decoupled during the design process. The fist cavity is suspended above the THz cavity (e.g. 1 μm above) and near the field maximum to ensure that the THz mode can effectively extent into the cavity without affecting the telecom-band modes. Therefore, there has existed a high overlap between the three fundamental cavity modes raised from the nonlinear susceptibility of the medium (e.g. GaAs, GaP) [198, 199]. Each cavity is over-coupled to a waveguide extending from one end, with the length of the Bragg mirror of holes at that end used to tune the strength of the coupling. The near-infrared pump and idler waves are input and the THz output wave is collected via these waveguide modes. This feature is schematically presented in Fig. 1.62a. The normalized mode profile of the terahertz radiation is sketched in Fig. 1.62b and the diagram of the near-infrared dual-mode cavity illustrating the spatial profile of the nonlinear mode-product ($E_{y,TE} \cdot E_{x,TM}$) is presented in Fig. 1.62c.

Because of performing the nonlinear process in the near-infrared cavity, the THz cavity can be composed of any material. This condition provides a degree of freedom. This point becomes much important by considering the scarcity of low-loss THz materials.

1.4.2.9 Silicon Waveguide Based Terahertz Source [200]

The basic problem in connection with terahertz generation techniques using second-order nonlinearities is having low conversion efficiencies [201, 202]. There are two reasons for having low conversion efficiencies: The first one is concerned with THz modes requiring large lateral dimensions, and second one is related to less efficient nonlinear frequency conversion at lower frequencies. Therefore using

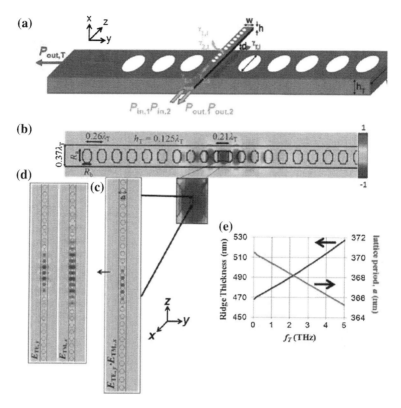

Fig. 1.62 **a** Schematic of the triply resonant system of coupled photonic crystal cavities for efficient THz generation. **b** Normalized mode profile of the THz mode. **c** Diagram of the near-infrared dual-mode cavity. **d** TE- and TM-like mode profiles of the near-infrared cavity. **e** Nanobeam thickness (*black*) and lattice period (*red*) for the near infrared PhCNC versus THz difference frequency [186]

pulsed lasers are suggested. In practice to produce continuous THz radiation by DFG technique, the following design factors should be considered: guiding optical modes with low loss in small, high-index-contrast, ridge waveguides, having polymers with high nonlinearities, the proper THz waveguide, and high power laser [203, 204, L.R. Dalton, Chemistry Department, University of Washington (2006, Personal communication)].

A continuously tunable, milliwatt output, continuous wave, and room temperature source in the 0.5–14 THz regime based on a silicon waveguide for near-infrared radiation, contained within a metal waveguide for terahertz radiation was reported [205]. Using a nonlinear polymer cladding two near-infrared lasers is mixed and through difference-frequency generation one can produce terahertz output. The optical fields and nonlinear effect can be enhanced by having small structural dimensions. There are three challenges of designing a silicon terahertz source: designing waveguides for both optical and terahertz radiation without losses, using nonlinear materials in such way to optimize the modal overlap, and

Fig. 1.63 Schematic diagram of the terahertz source. The *narrow rectangle* corresponds to the silicon waveguide atop an oxide pillar, while the base corresponds to bulk silicon that has been micromachined. The *thick rectangles* indicate the metal waveguide structure, made of copper chosen for optimal performance [200]

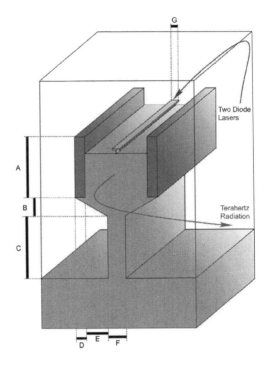

finally implementing the waveguides so that the phase-matching conditions are obtained between the THz and near-IR (NIR) signals.

It should be noted that while the silicon based waveguides can guide NIR radiation but they are transparent for THz radiation [206]. Also, due to silicon's high index, silicon waveguides can evanescently couple optical modes to nonlinear optical polymers [207]. On the other hand, available high-power laser sources in the range of 1480–1600 nm and also erbium-doped fiber amplifiers allow us to produce radiation ranges from dc to 15 THz with a continuous power output of 0.5 W [208]. Figure 1.63 shows a silicon terahertz generator geometry consisting of a silicon core for guiding NIR light, within a metal waveguide for confining THz radiation and various dimensions can be optimized in order to achieve phase matching.

1.5 Theoretically Investigation of Terahertz-Pulse Generation and Detection [1]

Different methods have been developed for generating THz pulses such as irradiation of photoconductive (PC) antennas, semiconductor surfaces, or quantum structures with femtosecond optical pulses. Also, in this section different methods for detection of a THz pulse are explained together with the emission of a THz pulse. These methods include variety of techniques from standard THz detection and generation to broadband and ultra broadband pulse generation and detection.

1.5.1 THz Pulse Generation and Detection Using Photoconductive Antenna (PC) [209]

Photoconductive antenna is one of the commonly used detectors and emitters for THz radiation which has a dipole-type antenna structure. The strip line antenna is fabricated on a PC substrate such as GaAs with a gold or Au:Ge:Ni alloy for ohmic contact and has a gap at the center of the antenna. Figure 1.64 shows a schematic of a PC antenna structure [209] with related dimensions where the PC gap (D) is ~5–10 μm, antenna width, W, is ~10–20 μm and antenna length, L, is ~30–50 μm for a Hertzian dipole-type antenna [210, 211]. The PC gap is biased with a dc voltage and is illuminated with femtosecond laser pulses. As it is illustrated in Fig. 1.64, the coplanar transmission line is designed to be long enough (e.g. 10 mm) to prevent the reflection at the line end. By excitation of the PC gap, photo-excited carriers are generated in the PC substrate and are accelerated under the bias field. These photo-excited carriers form an ultra-short current pulse with decay time of carrier lifetime in the substrate.

This transient current J (t) generates ultra-short electromagnetic pulsed radiation (THz radiation) where the terahertz field amplitude is proportional to the time derivative of photocurrent,

$$E_{THz} \propto \frac{\partial J(t)}{\partial t} \quad (1.9)$$

The peak value of the photocurrent is proportional to the averaged photocurrent, $\bar{J}$, divided by the duty ratio of the current pulse, T_{int}/τ_c, where T_{int} is the interval of the pump laser pulses and τ_c is the photo-carrier lifetime. By considering $\bar{G}$ as the time-averaged photo-conductance of PC gap, δ as the absorption depth of the pump light, V_b as the bias voltage, $\bar{\sigma}$ as the time-averaged conductivity, $\bar{n}_e$ as

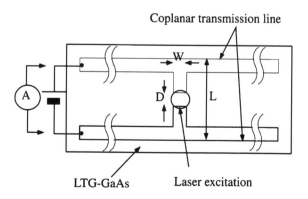

Fig. 1.64 Schematic of a PC antenna structure [209]

the averaged photo-carrier density, $h\nu$ as the photon energy of the pump laser, μ as the mobility of the carriers, P_{in} as the averaged pump laser power, R as the reflectance of the PC substrate and D as the size of PC gap, the peak value of the THz field can be described by:

$$\begin{aligned}E_{\text{THz}}^{\text{peak}} \propto \Delta J &\cong \bar{J}\frac{T_{\text{int}}}{\tau_c} = \bar{G}V_b\frac{T_{\text{int}}}{\tau_c} = \bar{\sigma}\frac{W\delta}{D}V_b\frac{T_{\text{int}}}{\tau_c} \\ &= e\mu\bar{n}_e\frac{W\delta}{D}V_b\frac{T_{\text{int}}}{\tau_c} = e\mu\tau_c\frac{(1-R)}{h\nu}\frac{P_{in}}{DW\delta}\frac{W\delta}{D}V_b\frac{T_{\text{int}}}{\tau_c} \\ &= e\mu T_{\text{int}}\frac{(1-R)}{h\nu}\frac{P_{in}}{D}\frac{V_b}{D}\end{aligned} \quad (1.10)$$

It is obvious from the above equation that the emission efficiency of a PC antenna is proportional to the carrier mobility but does not strongly depend on the carrier lifetime. Also, high carrier mobility and high applied bias voltage lead to higher efficiency although a high resistivity substrate is required to achieve to higher applicable bias voltage. Comparing the last two requirements it should be noted that satisfying high resistivity condition and acceptable carrier mobility is more feasible than high mobility condition with good resistivity. LTG GaAs (annealed) has an extraordinary high resistivity and a reasonably good mobility (100–300 cm^2 V^{-1} s^{-1}) and can satisfy the conditions. By assume the momentum relaxation time, τ_m, to be much shorter than the pulse width of the optical pump laser, τ_L, and the carrier lifetime, τ_c, ($\tau_m \ll \tau_L \ll \tau_c$) the carrier mobility can be considered constant and the carrier number, N (t), can be approximated by a time integral of the intensity profile of the pump laser I (t). Thus, the THz waveform from a PC emitter may be defined by:

$$E_{\text{THz}}^{\text{PC}}(t) \propto \frac{\partial J(t)}{\partial t} \propto \frac{\partial N(t)}{\partial t} \propto I(t) \quad (1.11)$$

The spectrum distribution of the THz field can then defined by the Fourier transform of the intensity profile of the pump laser:

$$E_{\text{THz}}^{\text{PC}}(\omega) \propto I(\omega) \quad (1.12)$$

It has been concluded that the bandwidth of the PC antenna is not limited by the slow carrier lifetime (~ 0.5 ps for LTG GaAs) nor by the carrier momentum relaxation time ($\sim$ several femtoseconds in LTG GaAs). However, the laser pulse width is the main parameter to enhance the broadband dynamics of the generation and detection of THz radiation with PC antennas. Thus, the bandwidth of a PC antenna can extend over 10 THz by using very short laser pulses. The fast Fourier transform (FFT) amplitude spectra of a PC antenna with dimensions L = 30 μm, D = 5 μm and W = 10 μm, excited by 20 fs laser pulses is depicted in Fig. 1.65. The THz radiation have generated from two different samples: a 12 μm thick ZnTe crystal and 0.1 mm thick GaSe crystal which was oriented 45° to the pump beam to attain a phase matching between the optical pump pulse and the THz pulse at about 25 THz which leads to increased efficiency and therefore an spectral peak

1.5 Theoretically Investigation of Terahertz-Pulse Generation and Detection

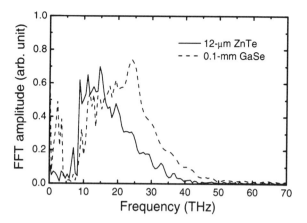

Fig. 1.65 FFT spectra of a PC dipole antenna for THz radiation generated from a ZnTe crystal (*solid curve*) and a GaSe crystal (*dashed curve*) [209]

at 25 THz. Spectral distributions of both samples over 30 THz displays the ultrabroadband property of the PC antenna detector.

1.5.2 THz Pulse Generation and Detection Using Other Methods [212]

Optical rectification in nonlinear media, charge oscillations in semiconductor quantum structures, coherent excitation of polar optical phonons, surge current drive by the surface depletion field of semiconductors, modulation of super current in a biased high-Tc bridge, and NLTLs are other methods than the PC antenna that the generate THz pulses. Figure 1.66 schematically presents these mechanisms.

The optical rectification process is associated with difference-frequency generation in the broad frequency spectra of the short optical pulse. This process can lead to THz-pulse generation through a second order nonlinear process as depicted in Fig. 1.66a. During this mechanism, a train of visible or near-IR short pulses is focused on a second-order nonlinear materials which can be dielectric crystals, semiconductors or organic materials. Therefore, the nonlinear polarization $P(\omega)$ created by the incident fields $E(\omega)$ can be described as:

$$P(\omega) = \varepsilon_0 \chi^{(2)}(\omega = \omega_1 - \omega_2) E(\omega_1) E^*(\omega_2) \qquad (1.13)$$

The emitted THz field can be defined through the time-domain polarization (Fourier transformation of $P(\omega)$) as:

$$E_{\text{THz}}(t) \propto \frac{\partial^2 P(t)}{\partial t^2} \qquad (1.14)$$

Various materials can be the host material for generation of a THz pulse through optical rectification such as LiNbO3, LiTaO3, ZnTe, InP, GaAs, GaSe, etc. Lower emission power and broader spectrum are the main advantages of

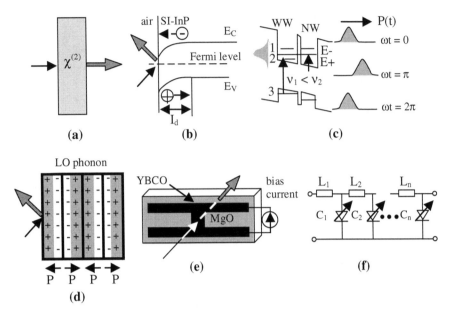

Fig. 1.66 THz-pulse emission mechanisms: **a** optical rectification in nonlinear media, **b** surge current and nonlinear process, **c** semiconductor Quantum structure, **d** coherent excitation of polar optical phonons, **e** high-Tc superconducting bridge, **f** NLTL [212]

optical rectification compared with PC antenna. The latter benefit arises from the instantaneous occurrence of this phenomenon. By taking into account the phase-matching condition, the emission efficiency of this process can be enhanced more.

The THz-pulse emission due to surface-depletion field effect is the second method depicted in Fig. 1.66b and this phenomenon can be observed in most wide-bandgap semiconductors. Donors and acceptors inside the crystal are trapped by the surface states and tend to form a charge-depletion region and thus a built-in surface electric field. In a p-type semiconductor with a band diagram sketched in Fig. 1.66b, electron–hole pairs are created due to absorption of photons with energies greater than the semiconductor band-gap and because of the built-in static field, electrons move to the surface and holes move to the wafer.

Therefore, a photocurrent which is called 'surge current' flows and a dipole layer builds up which emits THz pulses according to:

$$E(r,t) = \frac{l_e}{4\pi\varepsilon_0 c^2 r} \frac{\partial J(t)}{\partial t} \sin\theta \propto \frac{\partial J(t)}{\partial t} \qquad (1.15)$$

where $J(t)$ is the current in the dipole, l_e the effective length of the dipole, ε_0 the dielectric constant of vacuum, c the velocity of light in vacuum, and θ the angle from the direction of the dipole.

1.5 Theoretically Investigation of Terahertz-Pulse Generation and Detection

Excitonic charge oscillations in a single quantum well (SQW) or asymmetric double-coupled quantum well (DCQW) can lead to emission of THz radiation as schematically presented in Fig. 1.66c. This structure operates at a temperature of $\sim 10°K$. Applying an external electric field aligns the ground state electron energy levels of the wide and narrow wells and thus they are split into a bonding and an anti-bonding state ($|2\rangle$ and $|1\rangle$), with a splitting energy proportional to the tunneling probability of the carriers through the middle barrier. When these states are excited coherently with a broadband short optical pulse, charge oscillations that are due to quantum beats between these two levels occur with a beat frequency of $\Delta E/h$ ($\Delta E = E^- - E^+$). This charge oscillation results in a time-dependent polarization P(t) and hence dipole emission with a radiated electric field $E(t) \sim \partial^2 P(t)/\partial t^2$. A similar process in a SQW can be result in THz emission. However in this case the transitions between light-hole and heavy-hole exitons in an electric field are the mechanism of THz radiation generation.

Coherent excitation of Longitudinal Optical (LO) phonons with femtosecond optical pulses in different mediums such as semiconductors and semimetals can lead to generation of coherent phonons in the considered structure and the oscillation of the formed polarizations lead to THz emissions. This process is shown in Fig. 1.66d. The excitation of LO phonons is accomplished through ultrafast depolarization of the surface depletion field associated with the ultrafast polarization change within the surface-field region caused by the optical injection of carriers in a wide band-gap semiconductor such as GaAs.

Figure 1.66e illustrates another process for THz generation through the high-Tc superconducting bridge. The mechanism of this process is based on inverse process of the semiconductor PC switch in which the radiation is generated by optically short-circuiting the switch. A supercurrent flows through the bridge by cooling the device below the transition temperature and applying a bias current. Irradiation of the bridge with femtosecond optical pulse cause some Cooper pairs to break and change to quasi-particles instantaneously. Due to scattering phenomenon, these quasi-particles lose drift velocity and recombine into the Cooper pairs in a very short time before arriving next repetitively optical pulses. Thus, the supercurrent decreases quickly and recovers very rapidly (~ 1 ps).

The NLTL system is an all-electronic system consisted of a transmission line periodically loaded with reverse-biased Schottky varactor diodes which is used for generation and detection of electromagnetic transients as depicted in Fig. 1.66f. A wave travelling along the transmission line experiences a voltage-dependant propagation velocity due to existence of Schottky varactor diodes which act like voltage-variable capacitors. The voltage-dependant propagation velocity leads to formation of a shock wave (similar to a step function) which can be used as an ultrafast signal source and as a strobe generator for a diode sampling bridge. The output of the NLTL is coupled to a broadband bow-tie antenna as a transmitter with frequency-independent far-field radiation pattern and antenna impedance. The receiver is usually a bow-tie antenna interfaced to an NLTL-gated sampling circuit.

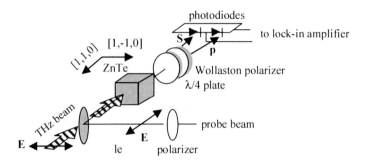

Fig. 1.67 Free-space EO sampling by ZnTe crystal [212]

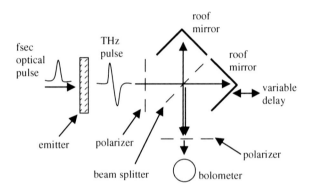

Fig. 1.68 Optical setup of a Martin–Puplett (M–P) type Fourier transform Interferometer [212]

Figure 1.67 shows the coherent detection of THz pulses based on the linear EO effect (Pockels effect) in an EO crystal. The refractive index of the EO crystal is modified in the presence of the THz pulse and a linearly polarized optical probe beam experiences this refractive index modification through its phase delay. The field strength of the THz pulse is proportional to the phase delay. A Wollaston polarizer converts the phase delay of the probe into intensity modulation and a pair of Si p-i-n photodiodes connected in a balanced mode is used to detect the optical intensity modulation. The difference signal of p-i-n photodiodes is fed to a lock-in amplifier and by measuring the output signal from the lock-in amplifier as a function of the time-delay of the probe beam one can obtain the THz-pulse waveform.

Fourier transform interferometry, two-source THz optoelectronic interferometry and autocorrelation-type interferometry are another class of methods that can be categorized into interferometry-based THz pulse detection methods.

Figure 1.68 shows a single THz radiation source and a Fourier transform interferometer which is typically in Michelson and Martin-Puplett (MP) mode with a liquid helium-cooled bolometer where the THz radiation source is considered more like the ordinary thermal source than the single-cycle electric pulse emitter. This configuration is capable of obtaining THz power information over the wide spectral range. However, the phase information will be lost in this technique.

1.5.3 THz Radiation from Bulk Semiconductor Microcavities

A simple method for generation of THz electromagnetic wave is excitation of unbiased semiconductor surfaces by short laser pulses. The excitation pulses in an optical cavity can be absorbed efficiently during the round-trips which allow the realization of a high static field only with the built-in potential. Figure 1.69 shows schematic structure and the related energy-band diagram around the THz-wave-generating region of the microcavity. This structure consists of a 182.6 nm-thick undoped GaAs layer and a 50 nm-thick Si-doped GaAs layer which are placed between two distributed Bragg reflectors (alternating AlAs and $Al_{0.2}Ga_{0.8}$ layers). The whole structure forms a one-wavelength cavity. The THz waves radiate from the surge current created by the laser pulses in the undoped GaAs layer, where a static electric field is built in as in the case of the bare n-type GaAs substrate. In most III–V semiconductors, the Fermi level at the surface is pinned at a certain energetic position, while it lies around the conduction band edge in n-doped bulk region. Therefore, in order to obtain a built-in surface electric field in the undoped GaAs region, the Fermi-level pinning was intentionally introduced at the interface between the undoped GaAs layer and the air-side DBR, by interrupting the molecular-epitaxial growth and exposing the GaAs surface to air.

The potential profile of n-GaAs and semiconductor microcavity near the surface at the thermal equilibrium is presented in Fig. 1.70 where part (a) exhibits the case that the light absorption is efficient but the field is low due to thick depletion region and a thinner field region is required for obtaining a higher field. Part (b) shows the high field but unfavorable case for exciting larger number of carriers (inefficient absorption). Hence, there is a tradeoff between the field strength and the photocarrier number. In an optical cavity, in contrast, the excitation pulses can be absorbed efficiently during the round trips as it mentioned before, even though a

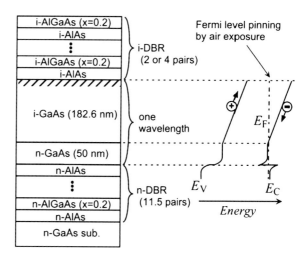

Fig. 1.69 Cross-sectional schematic and energy-band diagram around the THz-wave-generating region of the microcavity devices [213]

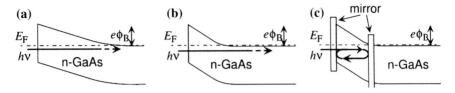

Fig. 1.70 Potential profile of n-GaAs and semiconductor microcavity near the surface at the thermal equilibrium. **a** Efficient light absorption, low field due to thick depletion region. **b** High field, inefficient light absorption due to thin depletion region. **c** High field and efficient absorption in semiconductor microcavity [213]

thin absorption layer is involved, allowing us to realize a rather high static field with only the built-in Potential as illustrated in Fig. 1.70c.

1.6 THz Detectors and Sources with Organic Materials [214]

In the past few years, serious attention has been paid to the use of organic materials to the THz generation and detection because of these materials have shown to contain great potential and broadened the material possibility. Organic electro-optic (EO) materials used in both the optoelectronic and all-optical techniques have made significant contribution to the recent development of all-optical THz systems. Lately it has been shown that the organic crystalline DAST (4-N, N-dimethylamino-4'-N'-methyl stilbazolium tosylate) and 2-(α-methylbeayl-amino)-5-nitropyridine (MBANP) to exhibit much higher EO coefficients in comparison with the inorganic materials which efficiently generate and detect the THz radiation [215–217]. More than that not only amorphous EO polymers show high EO coefficients but also the phonon absorption does not occur in these materials for THz range. Spectral gap-free bandwidth up to ~ 12 THz has been produced based on a polymer emitter-sensor pair [218]. The most interesting property of the EO tunability in polymers is achieved through the use of the composite constituent modification and film processing. The tunability property causes these materials to have excellent sensitivities, far great bandwidths and constant spectrum in the range between mid- and far-IR THz frequencies. Organic photoconductive dipole antennas (OPDAs) based on PPV and pentacene have also been successfully used for the propagation of the THz radiation. Next we review the organic materials which are used for THz sources and detectors [219, 220].

1.6.1 Conjugated Semiconducting Polymers

Photoconducting semiconductors are the bases for the construction of the photoconductive dipole antennas (PDAs)-based optoelectronic devices. Due to the study of Shirakawa et al. [221] on conductivity in polyacethylene, researchers have realized the existence of photoconductivity in conjugated semiconducting polymers. The idea is to have the possibility of easy and low cost fabrication of electronic and photonic devices with tunable properties. In particular, THz radiation from a poly (*p*-phenylene vinylene) (PPV) PDA and a pentecenen PDA have shown that the THz polymers photoconductors can be used in place of the THz inorganic photoconductors [219, 220]. The performance of polymer photoconductors is extremely depends on the physical parameters such as mobility and lifetime in the semiconducting polymers. Since the fundamental physics in connection with these is not clear, for instance, there is a disagreement on carrier photoexcitation picture in conjugated polymers, on the other hand there is also a dramatic difference in the magnitude of the measured mobility further efforts are needed for improving the device performance.

1.6.2 Organic EO Polymers

The spectral gap free, broadband with THz radiation is prevented by having the intrinsic lattice resonance phase-mismatch structure in inorganic EO materials. Instead the amorphous EO polymers do not have a lattice structure and have a quite flat refractive index ranging from NIR to FIR where this property provides obtaining a very good phase matching and hence broadband THz generation and detection. It should be noted that not only EO polymers have easy fabrication and low cost but also their tunability property has made them to be an excellent comparator to specific applications.

The selection of an active component-NLO chromophore along with a passive polymer matrix in a guest–host configuration are needed for design and building of an EO polymer [222]. In terms of applications the most important physical parameters associated with the guest NLO chromophore and host polymer are absorption spectra, dipole moments, second-order molecular optical polarizability and high glass transition temperature (T_g), film applicability, matching linear THz properties (absorption and refractive index) respectively. The addition of chromophore to polymer quantitatively will be such that the desire bulk EO coefficient is achieved without any phase separation. Since the inclusion of the chromophore molecules into the polymer matrix reduces T_g as a result the intrinsic T_g of the chosen polymer should be high enough. As a required fabrication step, electric field poling must be performed at or just below the T_g of the material where the NLO chromophore molecules may be easily oriented by the poling field and hence the centrosymmetric structure of chromophore will be broken to obtain the bulk EO coefficients. It should be noted that the composition of chromophore and polymer determines the electronic and linear optical properties not just chromophore alone. Also the matching capability of EO polymers to widely used lasers are important, for instant EO polymer consisting of an amorphous polycarbonate (APC) copolymer as the host and Lemke dye or DCDHF type dye (Fig. 1.71) as

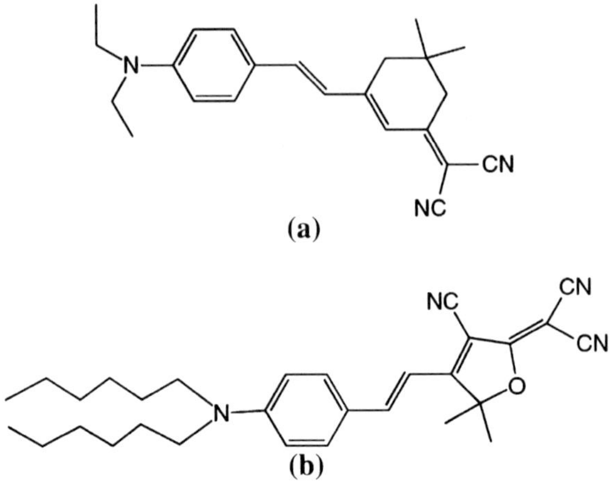

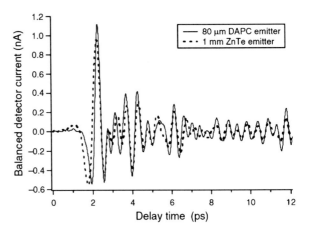

Fig. 1.71 Chemical structure of (**a**) Lemke and (**b**) DCDHF-6-V [214]

Fig. 1.72 The THz field emitted from a 1 mm thick ZnTe crystal and an 80 μm thick polymer layer (40%DCDHF-6-V/60%APC) using a 50-fs-duration, 800-nm-wavelength, 1-kHz, Ti:sapphire laser amplifying system [214]

the guest worked the best with Ti: sapphire lasers. The EO polymer is named based on Lemke as LAPC and that based on DCDHF as DAPC.

It is shown that EO polymers have coefficients one order of magnitude higher than ZnTe in THz emission range. For example Fig. 1.72 presents an 80 μm thick EO polymer film generating more THz emission than a 1000-μm-thick ZnTe crystal [223].

1.6.3 Organic EO Crystals

The most important organic EO crystal for THz applications is DAST which is a member of the Stilbazolium salt family and were grown by slow evaporation of saturated methanol solutions [215]. Despite its superior properties for non linear optical interactions, DAST has not been used practically because it is difficult to grow such large and high-quality single crystals till Mori et al. [224] proposed the slope nucleation method for growing high-quality DAST crystals. This technique combines spontaneous nucleation and the subsequent growth of a single crystal into one process. DAST crystals showed polarization dependent optical absorption in the visible range and strong birefringence of the low frequency (50 GHz– 2 THz) dielectric constants [215]. The optical absorption band extending furthest into the near infrared occurs for polarized light parallel to the 'a' crystal polar axis. Also strong optical absorption is observed for polarized light parallel to the 'b' crystal axis, with an absorption peak blue shifted by ∼50 nm relative to the 'a' axis. To study the dielectric properties, such as absorption and refractive index of DAST in the THz regime, THz time-domain spectroscopy and infrared spectroscopy were employed [225, 226]. Figure 1.73 shows the THz amplitude spectra from the DAST emitter-LAPC sensor pair when the 800-nm pump was a- and b-polarized, respectively [221].

For the THz-regime, the organic crystal DAST has been mostly proposed as nonlinear material because of its high nonlinear coefficient [225, 227]. Its suitability for THz-generation has been shown for difference-frequency generation (DFG) [228] and optical rectification [217, 229]. Monochromatic, ultra-wide tunability and ultra-broadband THz-wave generation with DAST crystal has been reported by using DFG, dual-wavelength KTiOPO$_4$ (KTP) optical parametric oscillator, and a femtosecond laser respectively [215, 228, 230, 231].

DAST-based THz-DFG technique which involves two pump beams with different wavelengths has the advantage of THz radiation being tuned by tuning the wavelength of one pump beam and in this way, a wide range of THz frequencies were achieved [228, 232]. For efficient DFG, the NLO-crystal is required to have large nonlinear, low-absorption coefficients. The generation of THz waves in the inorganic crystals with frequency above 10 THz is difficult due to strong phonon absorption in the crystals [233]. In return, the organic crystal DAST with a large nonlinear coefficient can produce ultra-wideband THz because of having low dielectric constant with an advantage for phasematching optical waves [227, 234]. THz-wave generation using difference-frequency mixing in DAST can be

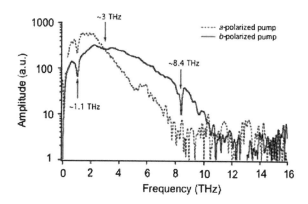

Fig. 1.73 THz amplitude spectra obtained from a DAST emitter—LAPC sensor pair operated at 800-nm wavelength with a- and b-polarization pump, respectively [214]

investigated theoretically as follow [233]. The THz output depends on the absorption of the crystal and the phase-matching conditions of DFG which are given by:

$$\text{energy conservation}: \left|\frac{1}{\lambda_1} - \frac{1}{\lambda_2}\right| = \frac{1}{\lambda_3} \tag{1.16}$$

$$\text{momentum conservation}: \left|\frac{n_1}{\lambda_1} - \frac{n_2}{\lambda_2}\right| = \frac{n_3}{\lambda_3} \tag{1.17}$$

where λ_1 and λ_2 are the input wavelengths, λ_3 is the DFG wavelength, n_1, n_2, and n_3 are the refractive indices at the respective wavelengths. The output power of the THz wave using DFG in the DAST crystal is given by [233, 235]:

$$P_3 = \frac{2\omega_3^2 d_{11}^2 L^2}{\varepsilon_0 c^3 n_1 n_2 n_3} \left(\frac{P_1 P_2}{\pi r^2}\right) T_1 T_2 T_3 S$$

$$S = \exp(-\alpha_3 L) \cdot \frac{1 + \exp(-\Delta\alpha L) - 2\exp\left(-\frac{1}{2}\Delta\alpha L\right)\cos(\Delta k L)}{(\Delta k L)^2 + \left(\frac{1}{2}\Delta\alpha L\right)^2}$$

$$\Delta k = k_1 - k_2 - k_3, \quad \Delta\alpha = |\alpha_1 - \alpha_2 - \alpha_3| \tag{1.18}$$

where P_1 and P_2 are the input peak powers and $P3$ is the peak THz power generated. L is the thickness of the DAST crystal. T_1, T_2 and T_3 are the Fresnel transmission coefficients, given by $T_j = 4n_j/(n_j + 1)^2, j = 1, 2, 3$, and Δk is the momentum mismatch. α_1, α_2 and α_3 are the absorption coefficients and ω_3 being the THz frequency. r is the radius of the beam focal spot. The refractive index n_3 and the absorption coefficient α_3 in the range of 0–3 THz were estimated using THz time-domain spectroscopy [225].

Figure 1.74 depicts the calculated THz-wave power both versus the input wavelength λ_1 (for generating 2.5- and 3-THz waves in a 1-mm-thick DAST crystal) and versus the DAST thickness (at 3 THz). Because of the strong absorption, the THz-wave power does not increase monotonically with the DAST thickness, but becomes saturated when the DAST crystal is thicker than 1 mm [233].

Figure 1.75 presents a schematic diagram of the experimental arrangement for DFG-based THz-wave generation in DAST crystal [233]. As a light source for THz-wave generation, a dual-wavelength optical parametric oscillator (OPO) consisting of two KTP crystals (in the same long cavity) and two flat mirrors (with high reflectance for the signal waves and high transmittance for the idler waves) was used. The signal wavelength was 840 to 900 nm, which corresponds to an idler wavelength of 1300 to 1450 nm. The pump source for the OPO was a diode-pumped, frequency-doubled, Q-switched Nd:YAG laser. The output beam of the OPO was focused on a spot on the DAST using a focal-length lens. The frequency of the THz wave was continuously tuned by changing the KTP crystal angle in the OPO cavity.

1.6 THz Detectors and Sources with Organic Materials

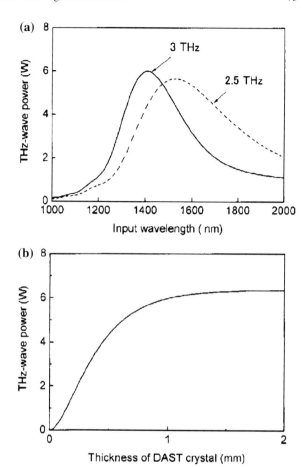

Fig. 1.74 The calculated THz-wave power versus (a) the input wavelength and (b) DAST thickness (at 3 THz, $\lambda_1 = 1450$ nm, $\lambda_2 = 1429$ nm, $n_1 = 2.14$, $n_1 = 2.142$) [233]

Figure 1.76 illustrates the THz-wave energy generated as a function of the THz frequency with 0.5- and 1-mm thick DAST crystals. Widely tunable THz waves from 2 to 20 THz were generated by mixing the output of an OPO over the 1300–1450 nm range [233]. The strong absorption induced by the resonance of the transverse optical phonon at 1.1 THz in DAST causes the wave energy to decrease below 2 THz. On the other hand, the decrease above 19 THz is due to the crystalline nature of the DAST crystal and to the strong absorption by the black polyethylene filter. The output energy at 2.5 THz for the 1-mm sample is about two times higher than that of the 0.5 mm sample. The reason for this is that the THz power in the low-loss region below 3 THz increases with the thickness of the DAST crystal, whereas due to saturation in the high-loss region above 3 THz, the output energy remains still as shown in Fig. 1.74.

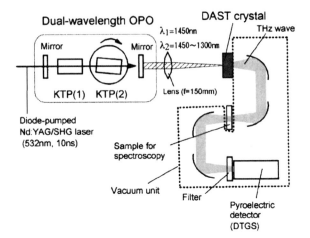

Fig. 1.75 Diagram of the experimental arrangement for DFG-based THz-wave generation in DAST crystal [233]

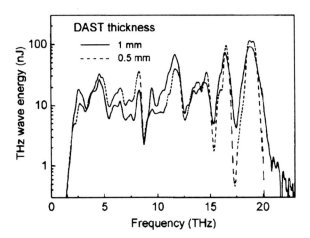

Fig. 1.76 THz output energy as a function of the THz frequency with 1- and 0.5-mm-thick DAST crystals [233]

THz pulse radiation from thin DAST crystals using optical rectification of a femtosecond laser pulse has been reported [215]. Due to the large second-order nonlinear susceptibility in DAST, the rectified field emitted from the organic salt is 185 and 42 times larger than LiTaO$_3$ and unbiased GaAs crystals respectively. In a conventional pump-probe arrangement, a continuous wave Ar laser pumped mode-locked Ti:sapphire laser was focused on the organic crystal. The optical gating for a photoconducting antenna was accomplished through the use of a weaker optical beam. A submillimeter-wave (rectified field) beam was detected from the organic crystal. A THz pulse radiation up to 15 THz (ultra-broadband) was demonstrated by pumping a 0.1-mm-thick DAST crystal using a 15-fs pulse laser [215].

The generation of THz radiation using the nonlinear interaction of two adjacent modes of a fiber amplifier also been done [236]. In this method, a passively

Q-switched $Cr^{4+}Nd^{3+}$:YAG laser is used to seed an Yb-fiber amplifier which is pumped by a P = 50 W laser diode emitting at a wavelength of 976 nm. The amplified radiation is collimated and focused into the nonlinear crystal and the generated radiation is then aligned with two parabolic mirrors and focused onto a bolometer. At the greater pump powers, side modes occur in the emission spectra of the fiber amplifier as shown in Fig. 1.77a, b. Difference frequency generation in the DAST crystal provides the emission of THz-radiation. Figure 1.77c shows the set-up of the laser system which is more compact as compared to the Ti:sapphire laser system.

Finally, it should be noted that another organic molecular crystal called MBANP was also used for the THz generation [217]. The material is highly absorptive for all wavelengths below 450 nm and exhibits an EO coefficient r_{eff} = 18.2 pm/V at 632.8 nm. Experimentally, generation of THz radiation using a ~200-µm-thick <001> MBANP crystal for 800 nm and 400 nm were accomplished based on optical rectification and direct intra-molecular charge transfer respectively.

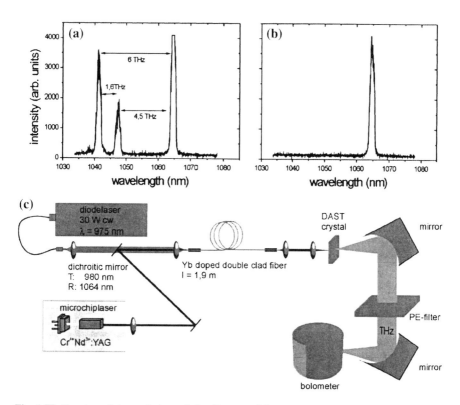

Fig. 1.77 Spectra of the emission of the fiber amplifier at a pump power: (**a**) P = 14.8 W and (**b**) P = 0 W (http://www.pi1.physik.uni-stuttgart.de/research/Methoden/THzMikroSpektrometer_e.php). **c** Set-up of the laser system for THz-generation [236]

1.6.4 Solutions Containing Polar Molecules

THz generation is always accompanied with the different types of charge transfer phenomena. Considering PDAs, inorganic, and organic EO media, there are three types of charge transfer phenomenon: the photogenerated carriers dynamics in the biased field, the deformation of the electron clouds of atoms and ion vibration, and the intramolecular electron transfer between donor and acceptor sites respectively [233]. Generating THz radiation from a Betaine-30 solution was carried out by Beard et al. [237]. To generate THz radiation in this solution, first a bias voltage is applied across the solution in order to orient the molecules and then ultra-short laser pulses are used to produce photo-induced charge transfer process where the transfer direction determines the polarity of the THz waveform. For solutions, poling order is not frozen and a real-time bias is necessary which leads to real-time poling with this having the advantage of shaping the THz waveform by varying the applied poling field at any time.

References

1. Kerecman, A.J.: The tungsten–P type silicon point contact diode. IEEE MTT-S Int. Microw. Symp. Dig. **73**, 30–34 (1973)
2. Fleming, J.W.: High resolution submillimeter-wave Fourier-transform spectrometry of gases. IEEE Trans. Microw. Theory Tech. **22**, 1023–1025 (1974)
3. Siegel, P.H.: Terahertz technology. Trans. Microw. Theory Tech. **50**(3), 910–928 (2002)
4. Semiconductor Physics Group, Department of Physics, University of Cambridge 2010. www.sp.phy.cam.ac.uk/.../WhatIsTHzImaging.htm.
5. Phillips, T.G., Keene, J.: Submillimeter astronomy. Proc. IEEE **80**, 1662–1678 (1992)
6. Leisawitz, D., Danchi, W.C., DiPirro, M.J., Feinberg, L.D., Gezari, D.Y., Hagopian, M., et al.: Scientific motivation and technology requirements for the SPIRIT and SPECS far-infrared/submillimeter space interferometers. Proc. SPIE **4013**, 36–46 (2000)
7. Waters, J.W.: Submillimeter-wavelength heterodyne spectroscopy and remote sensing of the upper atmosphere. Proc. IEEE **80**, 1679–1701 (1992)
8. Waters, J.W.: A 'Focused' MLS for EOS. Presented at the EOS Atmospheres Panel Presentation (1991)
9. Chyba, C.: Exploration of the solar system roadmap, science and mission strategy. NASA Office Space Sci. Solar Syst. Exploration Subcommittee, Galveston, TX (1999). http://solarsystem.nasa.gov/roadmap/pdffiles/Rmap.pdf
10. Gaidis, M.C.: Space-based applications of far infrared systems. In: 8th Int. Terahertz Electron. Conf., Darmstadt, Germany, pp. 125–128 (2000)
11. Costley, A.E., Hastie, R.J., Paul, J.W., Chamberlain, J.: Plasma temperature determination through electron resonance. Phys. Rev. Lett. **33**(13), 758–761 (1974)
12. Sattler, S., Hartfuß, H.J., W7-AS Team: Experimental evidence for electron temperature fluctuations in the core plasma of the W7-AS stellarator. Phys. Rev. Lett. **72**(5), 653–656 (1994)
13. DeLucia, F., Albert, S.: Fast-scanning spectroscopic method for the submillimeter: The FASSST spectrometer. In: Proc. SPIE Millimeter Submillimeter Waves Applicat. IV Conf., San Diego, CA, vol. 3465, pp. 236–246 (1998)

14. Jacobsen, R.H., Mittleman, D.M., Nuss, M.C.: Chemical recognition of gases and gas mixtures with terahertz waves. Opt. Lett. **21**(24), 2011–2013 (1996)
15. Petkie, D.T., Goyette, T.M., Bettens, R.P.A., Belov, S.P., Albert, S., Helminger, P., De Lucia, F.C.: A fast scan submillimeter spectroscopic technique. Rev. Sci. Instrum. **68**(4), 1675–1683 (1997)
16. Woolard, D., Kaul, R., Suenram, R., Walker, A.H., Globus, T., Samuels, A.: Terahertz electronics for chemical and biological warfare agent detection. In: IEEE MTT-S Int. Microwave Symp. Dig., Anaheim, CA, pp. 925–928 (1999)
17. Woolard, D., et al.: The potential use of submillimeter-wave spectroscopy as a technique for biological warfare agent detection. In: 22nd Army Sci. Conf., Baltimore, MD (2000)
18. Dexheimer, S.L. (Ed.): Terahertz Spectroscopy Principles and Applications, CRC Press (2008)
19. Izatt, J.R., Russell, B., Gagnon, R., Drouin, B.: Simultaneous measurement of moisture content and basis weight of paper sheet with a submillimeter laser. U.S. Patent 4 755 678 (1988)
20. Cantor, A.J., Cheo, P., Foster, M., Newman, L.: Application of submillimeter wave lasers to high voltage cable inspection. IEEE J. Quantum Electron. **17**, 477–489 (1981)
21. Woolard, D. Terahertz electronic research for defense: novel technology and science. In: 11th Int. Space Terahertz Tech. Symp., Ann Arbor, MI, pp. 22–38 (2000)
22. Brown, E.R.: All solid-state photomixing THz transmitter. Univ. California at Los Angeles, Los Angeles, CA. DARPA proposal to BAA 99-15 (1999)
23. Waldman, J., Fetterman, H.R., Duffy, P.E., Bryant, T.G., Tannenwald, P.E.: Submillimeter model measurements and their applications to millimeter radar systems. In: Proc. 4th Int. Infrared Near-MillimeterWaves Conf., pp. 49–50 (1979)
24. Coulombe, M.J., Horgan, T., Waldman, J., Szatkowski, G., Nixon, W.: A 524 GHz polarimetric compact range for scale model RCS measurements. In: Proc. Antenna Meas. Tech. Assoc., Monterey, CA (1999)
25. Griebel, M., Ospald, F., Smet, J.: Nanomaterials and systems for generating and detecting terahertz radiation. http://www.fkf.mpg.de/klitzing/research_topics/research_topics_details.php?topic=Nanomaterials%20and%20systems%20for%20generating%20and%20detecting%20terahertz%20radiation
26. Arnone, D.D., et al.: Applications of terahertz (THz) technology to medical imaging. In: Proc. SPIE Terahertz Spectroscopy Applicat. II (Munich, Germany), vol. 3823, pp. 209–219 (1999)
27. Markelz, A., et al.: Pulsed terahertz spectroscopy of DNA, bovine serum albumin and collagen between 0.1 and 2.0 THz. Chem. Phys. Lett. **320**, 42–48 (2000)
28. Sherwin, M.S., Schmuttenmaer, C.A., Bucksbaum, P.H. (eds.): Opportunities in THz science. Report of a DOE-NSF-NIH Workshop, Arlington, VA (2004)
29. Hundertmark, D.: A short introduction to Anderson localization. http://www.math.uiuc.edu/~dirk/preprints/localization3.pdf (2007)
30. Wiersma, D.: Laser physics: the smallest random laser. Nature (London) **406**, 132–133 (2000)
31. Alexander, S.B.: Optical Communication Receiver Design, pp. 129–132. SPIE Opt. Eng. Press/IEE, London (1997)
32. Samoska, L., et al.: InP MMIC chip set for power sources covering 80–170 GHz. Presented at the 12th Int. Space Terahertz Technol. Symp., San Diego, CA, 11.1 (2001)
33. Razeghi, M.: Technology of Quantum Devices. Springer, New York (2010)
34. Weinreb, S., Kerr, A.R.: Cryogenic cooling of mixers for millimeter and centimeter wavelengths. IEEE J. Solid-State Circuits **8**, 58–63 (1973)
35. McColl, M., Millea, M.F., Silver, A.H.: The superconductor–semiconductor Schottky barrier diode detector. Appl. Phys. Lett. **23**, 263–264 (1973)
36. Karpov, A., Miller, D., Rice, F., Zmuidzinas, J., Stern, J.A., Bumble, B., Leduc, H.G.: Lownoise 1.2 THz SIS receiver. In: 8th Int. Superconduct. Electron. Conf., Osaka, Japan, pp. 521–522 (2001)

37. McGrath, W.R.: Hot-electron bolometer mixers for submillimeter wavelengths: an overview of recent developments. In: 6th Int. Space Terahertz Technol. Symp., Pasadena, CA, pp. 216–228 (1995)
38. Skalare, A., McGrath, W.R., Echternach, P., LeDuc, H.G., Siddiqi, I., Verevkin, A., Prober, D.E.: Aluminum hot-electron bolometer mixers at submillimeter wavelengths. IEEE Trans. Appl. Superconduct. **11**, 641–644 (2001)
39. McGrath, W.R., Karasik, B.S., Skalare, A., Wyss, R., Bumble, B., LeDuc, H.G.: Hot-electron superconductive mixers for THz frequencies. In: SPIE Terahertz Spectroscopy Applicat. Conf., vol. 3617, pp. 80–88 (1999)
40. Karasik, B.S., Gaidis, M.C., McGrath, W.R., Bumble, B., LeDuc, H.G.: Lownoise in a diffusion-cooled hot-electron mixer at 2.5 THz. Appl. Phys. Lett. **71**, 1567–1569 (1997)
41. Wyss, R.A., Karasik, B.S., McGrath, W.R., Bumble, B., LeDuc, H.G.: Noise and bandwidth measurements of diffusion-cooled Nb hot-electron bolometer mixers at frequencies above the superconductive energy gap. In: 10th Int. Space Terahertz Technol. Symp., Charlottesville, VA, pp. 215–228 (1999)
42. Siegel P.H., Mehdi, I.: The spare-diode-detector: a new planardiode detector circuit with state-of-the-art performance for millimeter and submillimeter wavelengths. JPL, Pasadena, CA, JPL New Technol. Rep. NPO-20 104 (1996)
43. Erickson, N.R.: A fast and sensitive submillimeter waveguide power meter. In: 10th Int. Space Terahertz Technol. Symp., pp. 501–507 (1999)
44. Betz, A.L., Boreiko, R.T.: HgCdTe photoconductive mixers for 2–8 THz. In: 12th Int. Space Terahertz Technol. Conf., San Diego, CA, pp. 92–101 (2001)
45. Wang, N., et al.: Characterization of a submillimeter high-angular resolution camera with a monolithic silicon bolometer array for the CalTech submillimeter observatory. Appl. Opt. **35**(4), 6629–6640 (1996)
46. Mauskopf, P.D., Bock, J.J., Del Castillo, H., Holzapfel, W.L., Lange, A.E.: Composite infrared bolometers with Si N micromesh absorbers. Appl. Opt. **36**(4), 765–771 (1997)
47. Komiyama, S., Astafiev, O., Antonov, V., Hirai, H., Kutsuwa, T.: A single-photon detector in the far-infrared range. Nature **405**, 405–407 (2000)
48. Astafiev, O., Antonov, V., Kutsuwa, T., Komiyama, S.: Electrostatics of quantum dots in high magnetic fields studied by single-photon detection. Phys. Rev. B Condens. Matter **62**(24), R16731–R16743 (2000)
49. Schoelkopf, R.J., Wahlgren, P., Kozhevnikov, A.A., Delsing, P., Prober, D.E.: The radio-frequency single-electron transistor (RF-SET): a fast and ultrasensitive electrometer. Science **280**, 1238–1242 (1998)
50. Komiyama, S., Astafiev, O., Antonov, V., Kutsuwaa, T.: Single-photon detection of THz-waves using quantum dots. Microelectron. Eng. **63**, 173–178 (2002)
51. Byrd, J.: THz Detectors, Accelerator-Based Sources of Coherent Terahertz Radiation. UCSC, Santa Rosa (2008)
52. Courtesy of Thomas Keating Ltd and QMC Instruments Ltd (part of the Churchwood Trust Group of companies). http://qmciworks.ph.qmw.ac.uk/index.php?option=com_content&view=article&id=109&Itemid=533
53. Langley, S.P.: The bolometer. Nature (London) **25**, 14–16 (1881)
54. Shur, M.: Terahertz electronics. CS MANTECH Conference, Chicago, Illinois, USA (2008)
55. Shur, M.S.: Terahertz technology: devices and applications. In: Ghibaudo, G., Skotnicki, T., Cristoloveanu, S., Brillouet, M. (eds.) Proceedings of ESSDERC 2005, 35th European Solid-State Device Research Conference, Grenoble, France, pp. 13–21 (2005)
56. Han, P.Y., Zhang, X.-C.: Free-space coherent broadband terahertz time-domain spectroscopy. Meas. Sci. Technol. **12**, 1747–1756 (2001)
57. Smith, B.C.: Fourier Transform Infrared Spectroscopy. CRC Press, London (1996)
58. Kremer, F., Schönhals, A. (eds.): Broadband Dielectric Spectroscopy. Springer, New York (2002)
59. Crowe, T.W., Mattauch, R.J., Roser, H.P., Bishop, W.L., Peatman, W.C.B., Liu, X.: GaAs Schottky diodes for THz mixing applications. Proc. IEEE **80**(11), 1827–1841 (1992)

References

60. Shur, M.S., Eastman, L.F.: Ballistic transport in semiconductors at low-temperatures for low power high speed logic. IEEE Trans. Electron. Dev. **26**(11), 1677–1683 (1979)
61. Cooke, M.: Silicon transistor hits 500 GHz performance. III-Vs Review, 30–31 (2006)
62. Rieh, J.-S., Jagannathan, B., Greenberg, D.R., Meghelli, M., Rylyakov, A., Guarin, F., Yang, Z., Ahlgren, D.C., Freeman, G., Cottrell, P., Harame, D.: SiGe heterojunction bipolar transistors and circuits toward terahertz communication applications. IEEE Trans. Microw. Theory Tech. **52**, 2390–2408 (2004)
63. Snodgrass, W., Wu, B.-R., Cheng, K.Y., Feng, M.: Type-II GaAsSb/InP DHBTs with Record fT = 670 GHz and Simultaneous fT, fmax > 400 GHz. In: IEDM Technical Digest, pp. 663–666 (2007)
64. Lai, R., Mei, X. B., Deal, W.R., Yoshida, W., Kim, Y.M., Liu, P.H., Lee, J., et al.: A. Fung, Sub 50 nm InP HEMT Device with Fmax Greater than 1 THz. IEDM Technical Digest, pp. 609–611 (2007)
65. Lee, S., Jagannathan, B., Narasimha, S., Chou, A., Zamdmer, N., Johnson, J., Williams, R., Wagner, L., Kim, J., Plouchart, J.-O., Pekarik, J., Springer, S., Freeman, G.: Record RF performance of 45 nm SOI CMOS technology. In: IEDM Dig. Tech. Papers, pp. 255–258 (2007)
66. Knap, W., Teppe, F., Meziani, Y., Dyakonova, N., Lusakowski, J., Boeuf, F., Skotnicki, T., et al.: Plasma wave detection of sub-terahertz and terahertz radiation by silicon field-effect transistors. Appl. Phys. Lett. **85**, 675–677 (2004)
67. Teppe, F., Meziani, Y.M., Dyakonova, N., Lusakowski, J., Boeuf, F., Skotnicki, T., Maude, D., et al.: Terahertz detectors based on plasma oscillations in nanometric silicon field effect transistors. Physica Status Solidi C: Conferences **2**, 1413–1417 (2005)
68. Pala, N., Teppe, F., Veksler, D., Deng, Y., Shur, M.S., Gaska, R.: Nonresonant detection of terahertz radiation by silicon-on-insulator MOSFETs. Electron. Lett. **41**, 447–449 (2005)
69. Stillman, W., Shur, M.S., Veksler, D., Rumyantsev, S., Guarin, F.: Device loading effects on nonresonant detection of terahertz radiation by silicon MOSFETs. Electron. Lett. **43**, 422–423 (2007)
70. Stillman, W., Guarin, F., Kachorovskii, V.Y., Pala, N., Rumyantsev, S., Shur, M.S., Veksler, D.: Nanometer scale complementary silicon MOSFETs as detectors of terahertz and sub-terahertz radiation. In: 6th Annual IEEE Conference on Sensors, Atlanta, GA (2007)
71. McPheron, B.D.: Process development for a traveling wave terahertz detector, process & characterization. In: The NNIN REU Research Accomplishments (2009)
72. Van Zeghbroeck, B.: Principles of semiconductor devices. Section 2.3 (2007)
73. Bean, J., Tiwari, B., Szakmany, G., Bernstein, G., Fay, P., Porod, W.: Long wave infrared detection using dipole antenna-coupled metal-oxide-metal diodes. IEEE IRMMW-THz Digest (2008)
74. El Fatimy, A., Tombet, S.B., Teppe, F., Knap, W., Veksler, D.B., Rumyantsev, S., Shur, M.S., et al.: Terahertz detection by GaN/AlGaN transistors. Electron. Lett. **42**, 1342–1343 (2006)
75. Cao, Y., Jena, D.: High-mobility window for two-dimensional electron gases at ultrathin AlN/GaN heterojunctions. Appl. Phys. Lett. **90**, 182112-1–182112-3 (2007)
76. Zimmermann, T., Cao, Y., Deen, D., Simon, J., Fay, P., Jena, D., Xing, H.: AlN/GaN insulated gate HEMTs with 2.3 A/mm output current and 480 mS/mm transconductance. IEEE Electron. Dev. Lett. **29**(7), 661–664 (2008)
77. Cooke, M.: Pushing semiconductor detectors into the terahertz gap. III-Vs Rev. Adv. Semicond. Mag. **19**(8), 36–38 (2006)
78. In research being conducted by Andrea Markelz and Jonathan Bird at the University at Buffalo, 2006. http://electron.physics.buffalo.edu/spectre/
79. Bhattacharya, P., Huang, G., Yang, J.: Quantum dot photodetectors for THz detection, solid-state devices, theory, and technologies. In: Integrated Photonics and Optoelectronics, Research at the Solid State Electronics Laboratory (SSEL) (2010). http://www.mnf.umich.edu/SSEL/Projects/index.aspx?mid=3&sid=3&pid=47

80. Popa-Simil, L., Popa-Simil, I.L.: Multiband THz detection and imaging devices. Nanotech Conference Boston (2008). http://www.nsti.org/Nanotech2008/
81. Kawano, Y., Fuse, T., Toyokawa, S., Uchida, T., Ishibashi, K.: Terahertz photon-assisted tunneling in carbon nanotube quantum dots. J. Appl. Phys. **103**, 034307-1–034307-4 (2008)
82. Kawano, Y., Ishibashi, K.: An on-chip near-field terahertz probe and detector. Nat. Photon. **2**, 618–621 (2008)
83. Tucker, J.R., Feldman, M.J.: Quantum detection at millimeter wavelengths. Rev. Mod. Phys. **57**, 1055–1113 (1985)
84. Kouwenhoven, L.P., Jauhar, S., Orenstein, J., McEuen, P.L., Nagamune, Y., Motohisa, J., Sakaki, H.: Observation of photon-assisted tunneling through a quantum dot. Phys. Rev. Lett. **73**, 3443–3446 (1994)
85. Oosterkamp, T.H., Kouwenhoven, L.P., Koolen, A.E.A., van der Vaart, N.C., Harmans, C.J.P.M.: Photon sidebands of the ground state and first excited state of a quantum dot. Phys. Rev. Lett. **78**, 1536–1539 (1997)
86. Oosterkamp, T.H., Fujisawa, T., van der Wiel, W.G., Ishibashi, K., Hijman, R.V., Tarucha, S., Kouwenhoven, L.P.: Microwave spectroscopy of a quantum-dot molecule. Nature (London) **395**, 873–876 (1998)
87. Kawano, Y., Fuse, T., Toyokawa, S., Uchida, T., Ishibashi, K.: Highly sensitive and frequency- tunable THz detector using carbon nanotube quantum dots. In: IRMMW-THz, 33rd International Conference on Infrared, Millimeter and Terahertz Waves, 1-1 (2008). ISBN: 978-1-4244-2119-0
88. Zeuner, S., Allen, S.J., Maranowski, K.D., Gossard, A.C.: Photon-assisted tunneling in GaAs/AlGaAs superlattices up to room temperature. Appl. Phys. Lett. **69**, 2689–2691 (1996)
89. Tsukamoto, T., Moriyama, S., Tsuya, D., Suzuki, M., Yamaguchi, T., Aoyagi, Y., Ishibashi, K.: Carbon nanotube quantum dots fabricated on a GaAs/AlGaAs two-dimensional electron gas substrate. J. Appl. Phys. **98**, 076106-1–076106-3 (2005)
90. Vandersypen, L.M.K., Elzerman, J.M., Schouten, R.N., Willems van Beveren, L.H., Hanson, R., Kouwenhoven, L.P.: Real-time detection of single-electron tunneling using a quantum point contact. Appl. Phys. Lett. **85**, 4394–4396 (2004)
91. Franklin, N.R., Wang, Q., Tombler, T.W., Javey, A., Shim, M., Dai, H.: Integration of suspended carbon nanotube arrays into electronic devices and electromechanical systems. Appl. Phys. Lett. **81**, 913–915 (2002)
92. Kawano, Y., Okamoto, T.: Macroscopic channel-size effect of nonequilibrium electron distributions in quantum Hall conductors. Phys. Rev. Lett. **95**, 166801-1–166801-4 (2005)
93. Kawano, Y., Hisanaga, Y., Takenouchi, H., Komiyama, S.: Highly sensitive and tunable detection of far infrared radiation by quantum Hall devices. J. Appl. Phys. **89**, 4037–4048 (2001)
94. Kawano, Y., Ishibashi, K.: On-chip near-field terahertz detection based on a two-dimensional electron gas. Physica E **42**, 1188–1191 (2010)
95. Courtesy of RIKEN. http://www.optoiq.com/index/photonics-technologies-applications/lfw-display/lfw-article-display/365409/articles/laser-focus-world/volume-45/issue-7/features/terahertz-detectors-quantum-dots-enable-integrated-terahertz-imager.html
96. Yang, J., Agahi, F., Dai, D., Musante, C.F., Grammer, W., Lau, K.M., Yngvesson, K.S.: Wide-bandwidth electron bolometric mixers: a 2DEG prototype and potential for low-noise THz receivers. IEEE Trans. Microw. Theory Tech. **41**(4), 581–589 (1993)
97. Cherednichenko, S., Kroug, M., Khosropanah, P., Adam, A., Merkel, H., Kollberg, E., Loudkov, D., Voronov, B., Gol'tsman, G., Huebers, H.W., Richter, H.: 1.6 THz HEB mixer for far infrared space telescope (Hershel). Physica C **372**, 427–431 (2002)
98. Prober, D.: Superconducting terahertz mixer using a transition-edge microbolometer. Appl. Phys. Lett. **62**, 2119–2121 (1993)
99. Lee, M., Pfeiffer, L.N., West, K.W.: Ballistic cooling in a wideband two-dimensional electron gas bolometric mixer. Appl. Phys. Lett. **81**, 1243–1245 (2002)

100. Kong, J., Yenilmez, E., Wombler, T.W., Kim, W., Dai, H., Laughlin, R.B., Liu, L., Jayanthi, C.S., Wu, S.Y.: Quantum interference and ballistic transmission in nanotube electron waveguides. Phys. Rev. Lett. **87**, 106801-1–106801-4 (2001)
101. Santavicca, D.F., Prober, D.E.: Terahertz resonances and bolometric response of a single-walled carbon nanotubes. Paper 1646, 33rd Int. Conf. Infrared, Millimeter and Terahertz Waves, CalTech, Pasadena, CA (2008)
102. Yngvesson, K.S.: Very wide bandwidth hot electron bolometer heterodyne detectors based on single-walled carbon nanotubes. Appl. Phys. Lett. **87**, 043503-1–043503-3 (2005)
103. Carrion, E., Muthee, M., Donovan, J., Zannoni, R., Nicholson, J., Polizzi, E., Yngvesson, K.S.: New results on Terahertz detection by carbon nanotubes. In: Proc. 20th Int. Symp. Space THz Technology, Charlottesville (2009)
104. Kawase, K., Sato, M., Taniuchi, T., Ito, H.: Coherent tunable THz-wave generation from LiNbO with monolithic grating coupler. J. Appl. Phys. **68**(18), 2483–2485 (1996)
105. Matsuura, S., et al.: A traveling-wave THz photomixer based on angletuned phase matching. Appl. Phys. Lett. **74**(19), 2872–2874 (1999)
106. Noguchi, T., Ueda, A., Iwashita, H., Takano, S., Ishibashi, T., Ito, H., Nagatsuma, T.: Millimeter wave generation using a uni-traveling carrier photodiode. In: Presented at the 12th Int. Space Terahertz. Technol. Symp., San Diego, CA, Paper 3.2 (2001)
107. Xu, L., Zhang, X.-C., Auston, D.H.: Terahertz beam generation by femtosecond optical pulses in electro-optic materials. Appl. Phys. Lett. **61**(15), 1784–1786 (1992)
108. Mueller, E.R., Hesler, J., Crowe, T., Kurtz, D., Weikle, R.: Widelytunable laser sideband THz source for spectroscopy&LO applications. In: 12th Int. Space Terahertz. Technol. Symp., San Diego, CA, pp. 504–513 (2001)
109. Xu, B., Hu, Q., Melloch, M.R.: Electrically pumped tunable terahertz emitter based on intersubband transition. Appl. Phys. Lett. **71**(4), 440–442 (1997)
110. Sollner, T.C.L.G., Goodhue, W.D., Tannenwald, P.E., Parker, C.D., Peck, D.D.: Resonant tunneling through quantum wells at frequencies up to 2.5 THz. Appl. Phys. Lett. **43**, 588–590 (1993)
111. Reddy, M., Martin, S.C., Molnar, A.C., Muller, R.E., Smith, R.P., Siegel, P.H., Mondry, M.J., Rodwell, M.J. W., Allen, S.J. Jr.: Monolithic Schottky-collector resonant tunnel diode oscillator arrays to 650 GHz. In: 8th Int. Space Terahertz Technol. Symp., Cambridge, MA, pp. 149–161 (1997)
112. Song, I., Kang, K.-Y., Park, G.: Frequency-locked submillimeter wave generation from Josephson junction arrays. Jpn. J. Appl. Phys. pt. 1 **38**(1A), 44–47 (1999)
113. Weinreb, S., Gaier, T., Lai, R., Barsky, M., Leong, Y.C., Samoska, L.: High-gain 150–215 GHz MMIC amplifier with integral waveguide transitions. IEEE Microw. Guided Wave Lett. **9**, 282–284 (1999)
114. Samoska, L., Radisic, V., Micovic, M., Hu, M., Janke, P., Ngo, C.: InP MMIC chip set for power sources covering 80–170 GHz. In: Presented at the 12th Int. Space Terahertz Technol. Symp., San Diego, CA, Paper 11.1 (2001)
115. Faber, M.T., Chramiec, J., Adamski, M.E.: Microwave and millimeter-wave diode frequency multipliers. Artech House, Norwood (1995)
116. Maestrini, A., Bruston, J., Pukala, D., Martin, S., Mehdi, I.: Performance of a 1.2 THz frequency tripler using a GaAs frameless membrane monolithic circuit. In: IEEE MTT-S Int. Microwave Symp. Dig., Phoenix, AZ **3**, 1657–1660 (2001)
117. Maiwald, F., Martin, S., Bruston, J., Maestrini, A., Crawford, T., Siegel, P.H.: 2.7 THz tripler using monolithic membrane diodes. In: IEEE MTT-S Int. Microwave Symp. Dig., Phoenix, AZ, **3**, 1637–1640 (2001)
118. Miles, R.E., Garcia, J., Fletcher, J.R., Steenson, D.P., Chamberlain, J.M., Mann, C. M., Huq, E.J.: Modeling of micromachined klystrons for terahertz operation. In: 8th Int. Terahertz Electron. Conf., Darmstadt, Germany, pp. 55–58 (2000)
119. Siegel, P.H., Fung, A., Manohara, H., Xu, J., Chang, B.: Nanoklystron: A monolithic tube approach to terahertz power generation. In: 12th Int. Space Terahertz Technol. Symp., San Diego, CA, pp. 81–90 (2001)

120. East, J., Haddad, G.: Ballistic tunneling transit time device for terahertz power generation. In: 12th Int. Space Terahertz Technol. Symp., San Diego, CA, pp. 62–72 (2001)
121. Capasso, F., Colombelli, R., Paiella, R., Gmachl, C., Tredicucci, A., Sivco, D.L., Cho, A.Y.: Far-infrared and ultra-high-speed quantum-cascade lasers. Opt. Photon. News **12**(5), 40–46 (2001)
122. Nuss, M.C., Orenstein, J.: Terahertz time domain spectroscopy (THz-TDS). In: Gruener, G. (ed.) Millimeter-Wave Spectroscopy of Solids. Springer-Verlag, Berlin (1997)
123. Kawase, K., Sato, M., Taniuchi, T., Ito, H.: Coherent tunable terahertz-wave generation from LiNbO with monolithic grating coupler. Appl. Phys. Lett. **68**(18), 2483–2485 (1996)
124. Davies, A.G., Linfield, E.H., Johnston, M.B.: The development of terahertz sources and their applications. Phys. Med. Biol. **47**, 3679–3689 (2002)
125. Rochat, M., Ajili, L., Willenberg, H., Faist, J., Beere, H., Davies, G., Linfield, E., Ritchie, D.: Low-threshold terahertz quantum-cascade lasers. Appl. Phys. Lett. **81**(8), 1381–1383 (2002)
126. Mechold, L., Kunsh, J.: QCL modules are ready for industrial application. Laser Focus World **40**(5), 88–92 (2004)
127. Koehler, R., et al.: Terahertz semiconductor-heterostructure laser. Nature **417**, 156–159 (2002)
128. Kumar, S., et al.: Terahertz semiconductor heterostructure laser. Appl. Phys. Lett. **84**, 2494–2496 (2004)
129. Watanabe, Y., et al.: Component spatial pattern analysis of chemicals using terahertz spectroscopic imaging. Appl. Phys. Lett. **83**, 800–802 (2003)
130. Chamberlain, J.M., Miles, R.E., Collins, C.E., Steenson, D.P.: New directions in terahertz technology. In: Chamberlain, J.M., Miles, R.E. (eds.) NATO ASI Series. Kluwer, Dordrecht (1997)
131. Eisele, H., Rydberg, A., Haddad, G.I.: Recent advances in the performance of InP Gunn devices and GaAs TUNNETT diodes for the 100–300-GHz frequency range and above. IEEE Trans. Microw Theory Tech. **48**, 626–631 (2000)
132. Faist, J., Capasso, F., Sivco, D.L., Sirtori, C., Hutchinson, A.L., Cho, A.Y.: Quantum cascade laser. Science **264**, 553–556 (1994)
133. Colombelli, R., Capasso, F., Gmachl, C., Hutchinson, A.L., Sivco, D.L., Tredicucci, A., Wanke, M.C., et al.: Far-infrared surface-plasmon quantum-cascade lasers at 21.5 μm and 24 μm wavelengths. Appl. Phys. Lett. **78**, 2620–2622 (2001)
134. Hu, B.B., Nuss, M.C.: Imaging with terahertz waves. Opt. Lett. **20**, 1716–1718 (1995)
135. Nuss, M.C., Orenstein, J.: In: Grüner, G. (ed.) Millimeter and Submillimeter Wave Spectroscopy of Solids. Springer, Berlin (1998)
136. Wu, Q., Litz, M., Zhang, X.-C.: Broadband detection capability of ZnTe electro-optic field detectors. Appl. Phys. Lett. **68**, 2924–2926 (1996)
137. Han, P.Y., Tani, M., Pan, F., Zhang, X.-C.: Use of the organic crystal DAST for terahertz beam applications. Opt. Lett. **25**, 675–677 (2000)
138. Ma, X.F., Zhang, X.-C.: Determination of ratios between nonlinear-optical coefficients by using subpicosecond optical rectification. J. Opt. Soc. Am. B **10**, 1175–1179 (1993)
139. Seeta, P.N., Greene, B.I., Chuang, S.L.: Short terahertz pulses from semiconductor surfaces: the importance of bulk difference-frequency mixing. Appl. Phys. Lett. **63**, 3482–3484 (1993)
140. Rice, A., Jin, Y., Ma, X.F., Zhang, X.-C.: Terahertz optical rectification from <110> zinc-blende crystals. Appl. Phys. Lett. **64**, 1324–1326 (1994)
141. Corchia, A., Ciesla, C.M., Arnone, D.D., Linfield, E.H., Simmons, M.Y., Pepper, M.: Crystallographic orientation dependence of bulk optical rectification. J. Mod. Opt. **47**, 1837–1845 (2000)
142. Auston, D.H.: Picosecond optoelectronic switching and gating in silicon. Appl. Phys. Lett. **26**, 101–103 (1975)
143. Grischkowsky, D.R.: Optoelectronic characterization of transmission lines and waveguides by terahertz time-domain spectroscopy. IEEE J. Sel. Topics Quantum Electron. **6**, 1122–1135 (2000)

144. Zhang, X.-C., Hu, B.B., Darrow, J.T., Auston, D.H.: Generation of femtosecond electromagnetic pulses from semiconductor surfaces. Appl. Phys. Lett. **56**, 1011–1013 (1990)
145. Hu, B.B., Zhang, X.-C., Auston, D.H.: Terahertz radiation induced by subband-gap femtosecond optical excitation of GaAs. Phys. Rev. Lett. **67**, 2709–2712 (1991)
146. Gu, P., Tani, M., Kono, S., Sakai, K., Zhang, X.-C.: Study of terahertz radiation from InAs and InSb. J. Appl. Phys. **91**(9), 5533–5537 (2002)
147. Kono, S., Gu, P., Tani, M., Sakai, K.: Temperature dependence of terahertz radiation from n-type InSb and n-type InAs surfaces. Appl. Phys. B (Laser and Optics) **71**, 901–904 (2000)
148. Weiss, C., Wallenstein, R., Beigang, R.: Magnetic-field-enhanced generation of terahertz radiation in semiconductor surfaces. Appl. Phys. Lett. **77**, 4160–4162 (2000)
149. McLaughlin, R., Corchia, A., Johnston, M.B., Chen, Q., Ciesla, C.M., Arnone, D.D., Jones, G.A.C., et al.: Enhanced coherent terahertz emission from indium arsenide in the presence of a magnetic field. Appl. Phys. Lett. **76**, 2038–2040 (2000)
150. Corchia, A., McLaughlin, R., Johnston, M.B., Whittaker, D.M., Arnone, D.D., Linfield, E.H., Davies, A.G., Pepper, M.: Effects of magnetic field and optical fluence on terahertz emission in gallium arsenide. Phys. Rev. B **64**, 205204-1–205204-5 (2001)
151. Heyman, J.N., Neocleous, P., Hebert, D., Crowell, P.A., Müller, T., Unterrainer, K.: Terahertz emission from GaAs and InAs in a magnetic field. Phys. Rev. B **64**, 085202-1–085202-7 (2001)
152. Johnston, M.B., Corchia, A., Dowd, A., Linfield, E.H., Davies, A.G., McLaughlin, R., Arnone, D.D., Pepper, M.: Magnetic-field-induced enhancement of terahertz emission from III–V semiconductor surface. Physica E **13**, 896–899 (2001)
153. Han, P.Y., Cho, G.C., Zhang, X.-C.: Time-domain transillumination of biological tissues with terahertz pulses. Opt. Lett. **25**, 242–244 (2000)
154. Leitenstorfer, A., Hunsche, S., Shah, J., Nuss, M.C., Knox, W.H.: Femtosecond high-field transport in compound semiconductor. Phys. Rev. B **61**, 16642–16652 (2000)
155. Maestrini, A., Ward, J., Chattopadhyay, G., Schlecht, E., Mehdi, I.: Terahertz sources based on frequency multiplication and their applications. J. RF Eng. Telecommun. **62**(5/6), 118–122 (2008)
156. Ward, J., Schlecht, E., Chattopadhyay, G., Maestrini, Gill, J., Maiwald, F., Javadi, H., Mehdi, I.: Capability of THz sources based on Schottky diode frequency multiplier chains. IEEE MTT-S Digest, pp. 1587–1590 (2004). doi: 0-7803-8331-1/04
157. Maestrini, A., Ward, J., Gill, J., Javadi, H., Schlecht, E., Chattopadhyay, G., Maiwald, F., Erickson, N.R., Mehdi, I.: A 1.7 to 1.9 THz local oscillator source. IEEE Microw. Wirel. Compon. Lett. **14**(6), 253–255 (2004)
158. Maestrini, A., Ward, J.S., Javadi, H., Tripon-Canseliet, C., Gill, J., Chattopadhyay, G., Schlecht, E., Mehdi, I.: Local oscillator chain for 1.55 to 1.75 THz with 100 μW peak power. IEEE Microw. Wirel. Compon. Lett. **15**(12), 871–873 (2005)
159. Erickson, N.: Diode frequency multipliers for THz local oscillator applications. SPIE Conference on Advanced Technology MMW, Radio, and Terahertz Telescopes, SPIE **3357**, 75–84 (1998)
160. Maiwald, F., Schlecht, E., Ward, J., Lin, R., Leon, R., Pearson, J., Mehdi, I.: Design and operational considerations for robust planar GaAs varactors: a reliability study. In: Proceedings of 14th International Symposium on Space Terahertz Technology, Tucson, AZ, pp. 488–491 (2003)
161. Maestrini, A., Ward, J., Gill, J., Javadi, H., Schlecht, E., Tripon-Canseliet, C., Chattopadhyay, G., Mehdi, I.: A 540–640 GHz high efficiency four anode frequency tripler. IEEE Trans. Microw. Theory Tech. **53**, 2835–2843 (2005)
162. Maestrini, A., Tripon-Canseliet, C., Ward, J.S., Gill, J.J., Mehdi, I.: A high efficiency multiple-anode 260–340 GHz frequency tripler. In: Proceedings of the 17th International Conference on Space Terahertz Technology, Paris, paper P2-05, pp. 233–236 (2006)
163. Gallerano, G.P., Biedron, S.: Overview of terahertz radiation sources. In: Proceedings of the FEL Conference, pp. 216–221 (2004)

164. Staprans, A., McCune, E., Ruetz, J.: High-power linear-beam tubes. Proc. IEEE **61**, 299–330 (1973)
165. Stuart, R.A., Al-Shamma'a, A.I., Lucas, J.: Compact tuneable terahertz source. In: 2nd EMRS DTC Technical Conference, Edinburgh (2005)
166. Miles, R.E., Garcia, J., Fletcher, J.R., Steenson, D.P., Chamberlain, J.M., Mann, C.M., Huq, E.J.: Modeling of micromachined klystrons for terahertz operation. In: 8th Int. Terahertz Electron. Conf. Darmstadt, Germany, pp. 55–58 (2000)
167. Chamberlain, M., et al.: Miniaturized vacuum technologies: do they have a future for terahertz frequency devices? IEEE Conference, Pulsed power plasma science, PPPS-2001 Digest of technical papers **1**, 130–134 (2001)
168. Caryotakis, G., et al.: The klystrino: a high power W-band amplifier. In: International Vacuum Electronics Conference 2000 paper 1.5, Monterey, CA, USA (2000)
169. Bhattacharjee, S., et al.: Folded waveguide traveling-wave tube sources for terahertz radiation. IEEE Trans. Plasma Sci. **32**, 1002–1014 (2004)
170. Bjarnason, J.E., Chan, T.L.J., Lee, A.W.M., Brown, E.R., Driscoll, D.C., Hanson, M., Gossard, A.C., Muller, R.E.: ErAs:GaAs photomixer with two-decade tunability and 12 mW peak output power. Appl. Phys. Lett. **85**(18), 3983–3985 (2004)
171. Smith, F.W.: Ph.D. thesis. Massachusetts Institute of Technology (1990)
172. Kadow, C., Fleischer, S.B., Ibetson, J.P., Bowers, J.E., Gossard, A.C., Dong, J.W., Palmstrom, C.J.: Self-assembled ErAs islands in GaAs: growth and subpicosecond carrier dynamics. Appl. Phys. Lett. **75**, 3548–3550 (1999)
173. McIntosh, K.A., Brown, E.R., Nichols, K.B., McMahon, O.B., DiNatale, W.F., Lyszczarz, T.M.: Terahertz measurements of resonant planar antennas coupled to low-temperature-grown GaAs photomixers. Appl. Phys. Lett. **69**, 3632–3634 (1996)
174. Veselago, V.G.: The electrodynamics of substances with simultaneously negative values of epsilon and mu. Sov. Phys. Usp. **10**, 509–513 (1968)
175. Pendry, J.B., Holden, A.J., Robbins, D.J., Stewart, W.J.: Magnetism from conductors and enhanced nonlinear phenomena. IEEE Trans. Microw. Theory Tech. **47**, 2075–2084 (1999)
176. Pendry, J.B., Holden, A.J., Stewart, W.J., Youngs, I.: Extremely low frequency plasmons in metallic mesostructures. Phys. Rev. Lett. **76**, 4773–4776 (1996)
177. Mayes, M.G.: Miniature field deployable terahertz source. SPIE Defense and Security Conference, Orlando, FL (2006)
178. Marqués, R., Medina, F., Rafii-El-Idrissi, R.: Role of bianisotropy in negative permeability and left-handed metamaterials. Phys. Rev. B **65**, 144440-1–144440-6 (2002)
179. Padilla, W.J., Aronsson, M.T., Highstrete, C., Lee, M., Taylor, A.J., Averitt, R.D.: Electrically resonant terahertz metamaterials: theoretical and experimental investigations. Phys. Rev. B **75**, 041102-1–041102-4 (2007)
180. Hidaka, T., Matsuura, S., Tani, M., Sakai, K.: CW terahertz wave generation by photomixing using a two-longitudinal-mode laser diode. Electron. Lett. **33**, 2039–2040 (1997)
181. Ostmann, T.K., Knobloch, P., Koch, M., Hoffmann, S., Breede, M., Hofmann, M., Hein, G., Pierz, K., Sperling, M., Donhuijsen, K.: Continuous-wave THz imaging. Electron. Lett. **37**, 1461–1463 (2001)
182. Diehl, L., Bour, D., Corzine, S., Zhu, J., Höfler, G., Lonar, M., Troccoli, M., Capasso, F.: High-power quantum cascade lasers grown by low-pressure metal organic vapor-phase epitaxy operating in continuous wave above 400 K. Appl. Phys. Lett. **88**, 201115-1–201115-3 (2006)
183. Ragam, S., Tanabe, T., Saito, K., Oyama, Y., Nishizawa, J.: Enhancement of CW THz wave power under noncollinear phase-matching conditions in difference frequency generation. J. Lightwave Technol. **27**, 3057–3061 (2009)
184. Nishizawa, J., Tanabe, T., Suto, K., Watanabe, Y., Sasaki, T., Oyama, Y.: Continuous-wave frequency-tunable terahertz-wave generation from GaP. IEEE Photon. Technol. Lett. **18**, 2008–2010 (2006)

185. Sowade, R., Breunig, I., Mayorga, I.C., Kiessling, J., Tulea, C., Dierolf, V., Buse, K.: Continuous-wave optical parametric terahertz source. Opt. Express **17**(25), 22303–22310 (2009)
186. Burgess, I.B., Zhang, Y., McCutcheon, M.W., Rodriguez, A.W., Bravo-Abad, J., et al.: Design of an efficient terahertz source using triply resonant nonlinear photonic crystal cavities. Opt. Express **17**(22), 20099–20108 (2009)
187. Belkin, M.A., Capasso, F., Belyanin, A., Sivco, D.L., Cho, A.Y., Oakley, D.C., Vineis, C.J., Turner, G.W.: Terahertz quantum-cascade-laser source based on intracavity difference-frequency generation. Nat. Photon. **1**, 288–292 (2007)
188. Bieler, M.: THz generation from resonant excitation of semiconductor nanostructures: Investigation of secondorder nonlinear optical effects. IEEE J. Sel. Top. Quantum Electron. **14**, 458–469 (2008)
189. Andronico, A., Claudon, J., Gerard, J.M., Berger, V., Leo, G.: Integrated terahertz source based on three-wave mixing of whispering-gallery modes. Opt. Lett. **33**, 2416–2418 (2008)
190. Vodopyanov, K.L., Fejer, M.M., Yu, X., Harris, J.S., Lee, Y.S., Hurlbut, W.C., Kozlov, V.G., et al.: Terahertz-wave generation in quasi-phase-matched GaAs. Appl. Phys. Lett. **89**, 141119-1–141119-3 (2006)
191. Imeshev, G., Fermann, M.E., Vodopyanov, K.L., Fejer, M.M., Yu, X., Harris, J.S., Bliss, D., Lynch, C.: High-power source of THz radiation based on orientation-patterned GaAs pumped by a fiber laser. Opt. Express **14**, 4439–4444 (2006)
192. Hebling, J., Stepanov, A.G., Almassi, G., Bartal, B., Kuhl, J.: Tunable THz pulse generation by optical rectification of ultrashort laser pulses with tilted pulse fronts. Appl. Phys. B **78**, 593–599 (2004)
193. Van Exter, M., Grischkowsky, D.R.: Characterization of an optoelectronic terahertz beam system. IEEE Trans. Microw. Theory Tech. **38**, 1684–1691 (1990)
194. Wu, Q., Litz, M., Zhang, X.C.: Broadband detection capability of ZnTe electro-optic field detectors. Appl. Phys. Lett. **68**, 2924–2926 (1996)
195. Lee, Y.S., Meade, T., Perlin, V., Winful, H., Norris, T.B., Galvanauskas, A.: Generation of narrow-band terahertz radiation via optical rectification of femtosecond pulses in periodically poled lithium niobate. Appl. Phys. Lett. **76**, 2505–2507 (2000)
196. Schaar, J.E., Vodopyanov, K.L., Fejer, M.M.: Intracavity terahertz-wave generation in a synchronously pumped optical parametric oscillator using quasi-phase-matched GaAs. Opt. Lett. **32**, 1284–1286 (2007)
197. McCutcheon, M.W., Youmg, J.F., Reiger, G.W., Dalacu, D., Frederick, S., Poole, P.J., Wiliams, R.L.: Experimental demonstration of second-order processes in photonic crystal microcavities at submilliwatt excitation powers. Phys. Rev. B **76**, 245104-1–245104-6 (2007)
198. McCutcheon, M.W., Chang, D.E., Zhang, Y., Lukin, M.D., Lončar, M.: Broad-band spectral control of single photon sources using a nonlinear photonic crystal cavity. arXiv:0903.4706 (2009)
199. Singh, S.: Nonlinear optical materials. In: Weber, M.J. (ed.) Handbook of Laser Science and Technology, vol. III: Optical Materials, Part I. CRC Press (1986)
200. Baehr-Jones, T., Hochberg, M., Soref, R., Scherer, A.: Design of a tunable, room temperature, continuous-wave terahertz source and detector using silicon waveguides. J. Opt. Soc. Am. B **25**(2), 261–268 (2008)
201. Sasaki, Y., Yuri, A., Kawase, K., Ito, H.: Terahertz-wave surface-emitted difference frequency generation in slantstripe-type periodically poled LiNbO3 crystal. Appl. Phys. Lett. **81**, 3323–3325 (2002)
202. Kukushkin, V.: Efficient generation of terahertz pulses from single infrared beams in C/GaAs/C waveguiding heterostructures. J. Opt. Soc. Am. B **23**, 2528–2534 (2006)
203. Enami, Y., Derose, C.T., Mathine, D., Loychik, C., Greenlee, C., Norwood, R.A., Kim, T.D., et al.: Hybrid polymer/sol-gel waveguide modulators with exceptionally large electrooptic coefficients. Nat. Photon. **1**, 180–185 (2007)

204. Baehr-Jones, T., Hochberg, M., Wang, G.X., Lawson, R., Liao, Y., Sullivan, P.A., Dalton, L., et al.: Optical modulation and detection in slotted silicon waveguides. Opt. Express **13**, 5216–5226 (2005)
205. Mueller, E.: Terahertz radiation sources for imaging and sensing applications. Photonics Spectra **40**, 60–69 (2006)
206. Palik, E.: Handbook of Optical Constants of Solids. Academic (1985)
207. Hochberg, M., Baehr-Jones, T., Wang, G.X., Shearn, M., Harvard, K., Luo, J.D., Chen, B.Q., et al.: Terahertz all-optical modulation in a silicon-polymer hybrid system. Nat. Mater. **5**, 703–709 (2006)
208. Plant, J., Juodawlkis, P.W., Huang, R.K., Donnelly, J.P., Missaggia, L.J., Ray, K.G.: 1.5-μm InGaAsP-InP slabcoupled optical waveguide lasers. IEEE Photonics Technol. Lett. **17**, 735–737 (2005)
209. Tani, M., Herrmann, M., Sakai, K.: Generation and detection of terahertz pulsed radiation with photoconductive antennas and its application to imaging. Meas. Sci. Technol. **13**, 1739–1745 (2002)
210. Smith, P.R., Auston, D.H., Nuss, M.C.: Subpicosecond photoconducting dipole antennas. IEEE J. Quantum Electron. **24**, 255–260 (1988)
211. Tani, M., Matsuura, S., Sakai, K., Nakashima, S.: Emission characteristics of photoconductive antennas based on low-temperature-grown GaAs and semi-insulating GaAs. Appl. Opt. **36**, 7853–7859 (1997)
212. Sakai, K. (ed.): Terahertz Optoelectronics. Springer-Verlag, Berlin (2005)
213. Sakurada, T., Kadoya, Y., Yamanishi, M.: THz electromagnetic wave radiation from bulk semiconductor microcavities excited by short laser pulses. Jpn. J. Appl. Phys. **41**, L256–L259 (2002)
214. Zheng, X., McLaughlin, C.V., Cunningham, P., Michael Hayde, L.: Organic broadband terahertz sources and sensors. J. Nanoelectron. Optoelectron. **2**, 1–19 (2007)
215. Zhang, X.-C., Ma, X.F., Jin, Y., Lu, T.-M., Boden, E.P., Phelps, P.D., et al.: Terahertz optical rectification from a nonlinear organic crystal. Appl. Phys. Lett. **61**, 3080–3082 (1992)
216. Carrig, T.J., Rodriguez, G., Clement, T.S., Taylor, A.J., Stewart, K.R.: Scaling of terahertz radiation via optical rectification in electro-optic crystals. Appl. Phys. Lett. **66**, 121–123 (1995)
217. Carey, J.J., Bailey, R.T., Pugh, D., Sherwood, J.N., Cruickshank, F.R., Wynne, K.: Terahertz pulse generation in an organic crystal by optical rectification and resonant excitation of molecular charge transfer. Appl. Phys. Lett. **81**, 4335–4337 (2002)
218. Zheng, X., Sinyukov, A., Hayden, L.M.: Broadband and gap-free response of a terahertz system based on a poled polymer emitter-sensor pair. Appl. Phys. Lett. **87**, 081115-1–081115-3 (2005)
219. Soci, C., Moses, D.: Terahertz generation from poly (p-phenelene, vinylene) photoconductive antenna. Synth. Met. **139**, 815–817 (2003)
220. Ostroverkhova, O., Shcherbyna, S., Cooke, D.G., Egerton, R.F., Hegmann, F.A., Tykwinski, R.R., Parkin, S.R., Anthony, J.E.: Optical and transient photoconductive properties of pentacene and functionalized pentacene thin films: dependence on film morphology. J. Appl. Phys. **98**, 033701-1–033701-12 (2005)
221. Ito, T., Shirakawa, H., Ikeda, S.: Simultaneous polymerization and formation of polyacetylene film on the surface of concentrated soluble Ziegler-type catalyst solution. J. Polym. Sci. Chem. Ed. **12**, 11–20 (1974)
222. Hayden, L.M., Sinyukov, A.M., Leahy, M.R., French, J., Lindahl, P., Herman, W.N., Twieg, R.J., He, M.: New materials for optical rectification and electrooptic sampling ultrashort pulses in the terahertz regime. J. Polym. Sci. B: Polym. Phys. **41**, 2492–2500 (2003)
223. Sinyukov, A.M., Hayden, L.M.: Efficient electro-optic polymers for THz systems. J. Phys. Chem. B **108**, 8515–8522 (2004)
224. Mori, Y., Takahashi, Y., Iwai, T., Yoshimura, M., Yap, Y.K., Sasaki, T.: Slope nucleation method for the growth of high-quality 4-dmethylamino-methyl-4-stilbazolium-tosylate (DAST) crystals. Jpn. J. Appl. Phys. **39**, L1006–L1008 (2000)

225. Walther, M., KJensby, J., Keiding, S.R., Takahashi, H., Ito, H.: Far-infrared properties of DAST. Opt. Lett. **25**, 911–913 (2000)
226. Bosshard, C., Spreiter, R., De De giorgi, L., Gunter, P.: Infrared and Raman spectroscopy of the organic crystal polarization dependence and contribution of molecular vibrations DAST: to the linear electro-optic effect. Phys. Rev. B **66**, 205107(1)–205107(9) (2002)
227. Pan, F., Knöpfle, G., Bosshard, Ch., Follonier, S., Spreiter, R., Wong, M.S., Günter, P.: Electro-optic properties of the organic salt 4-N, N-dimethylamino-4'-N'-methyl-stilbazolium tosylate. Appl. Phys. Lett. **69**, 13–15 (1996)
228. Kawase, K., Mizuno, M., Sohma, S., Takahashi, H., Taniuchi, T., Urata, Y., Wada, S., et al.: Difference-frequency terahertz-wave generation from 4-N, N-dimethylamino-4'-N'-methyl-stilbazolium tosylate by use of an electronically tuned Ti: sapphire laser. Opt. Lett. **24**, 1065–1067 (1999)
229. Schneider, A., Biaggio, I., Günter, P.: Optimized generation of THz pulses via optical rectification in the organic salt DAST. Opt. Commun. **224**, 337–341 (2003)
230. Ito, H., Suizu, K., Yamashita, T., Nawahara, A., Sato, T.: Random frequency accessible broad tunable terahertz-wave source using phase-matched 4-dimethylamino-N-methyl-4-stilbazolium tosylate crystal. Jpn. J. Appl. Phys. **46**, 7321–7324 (2007)
231. Ashida, M., Akai, R., Shimosato, H., Katayama, I., Itoh, T., Miyamoto, K., Ito, H.: Ultrabroadband THz field detection beyond 170 THz with a photoconductive antenna. In: The Conference on Lasers and Electro-Optics (CLEO) Proceedings, CTuX6 (2008)
232. Kawase, K., Hatanaka, T., Takahashi, H., Nakamura, K., Taniuchi, T., Ito, H.: Tunable terahertz-wave generation from DAST crystal by dual signal-wave parametric oscillation of periodically poled lithium niobate. Opt. Lett. **25**, 1714–1716 (2000)
233. Taniuchi, T., Okada, S., Nakanishi, H.: Widely tunable terahertz-wave generation in an organic crystal and its spectroscopic application. J. Appl. Phys. **95**(11), 5984–5988 (2004)
234. Meier, U., Bosch, M., Bosshard, C., Pan, F., Gunter, P.: Parametric interactions in the organic salt 4-N, N-dimethylamino-4'-N'-methyl-stilbazolium tosylate at telecommunication wavelengths. J. Appl. Phys. **83**, 3486–3489 (1998)
235. Aggarwal, R.L., Lax, B.: Nonlinear Infrared Generation, p. 28. Springer, New York (1977)
236. Hohmann, K., Schippers, W., Willer, U., Bohling, C., Schade, W., Schossig, T.: Fiber-amplifier based THz source using difference-frequency generation in a DAST crystal. In: Quantum Electronics and Laser Science Conference (QELS) (2005)
237. Beard, M.C., Turner, G.M., Schmuttenmaer, C.A.: Measurement of electromagnetic radiation emitted during rapid intramolecular electron transfer. J. Am. Chem. Soc. **122**, 11541–11542 (2000)

Chapter 2
Terahertz and Infrared Quantum Photodetectors

2.1 Introduction

The importance and the application vastity of terahertz and infrared photodetectors are clear for every one working in this field. Infrared photodetectors are interesting components in optical communications, thermal imaging and sensor networking. Recently infrared photodetectors have been the focus of much attention due to its potential use in far-infrared imaging as well as room temperature operation, which is of interest from user's point of view. Conventionally all objects radiate most of its energy in the form of infrared and terahertz waves, for observing objects and physical activity in dark conditions one must monitor the infrared spectra. It is usually customary to use the 3–5 µm infrared window in military applications, 8–15 µm window in thermal imaging and >20 µm in THz applications such as medical diagnostics. More study is done on finding proper material for detecting the infrared spectrum. However the detection of long-wavelength infrared (e.g. 10 µm) radiation requires a small gap ($E_g \approx 0.1$ eV). Such small-bandgap materials are well-known to be more difficult to grow, process, and fabricate into devices than are larger bandgap semiconductors [1]. This problem is more critical in THz wavelengths in a manner that it is actually impossible to detect THz radiations via interband optical transitions. Intersubband transitions are a suitable alternative to cover the infrared and THz spectra. On these lines to remove the present problems, improve the detecting parameters and for integrating it with other optoelectronic devices, optimizing the trade-off between Responsivity and Detectivity, spectra engineering and obtaining to suitable detecting parameters at room temperature, photodetector structures are developed from bulk to quantum wells (QWIP) and dots (QDIP).

In this chapter we study different structures of QW and QD terahertz-infrared photodetectors. The main aim in design of all structures is to overcome the challenging subjects in long-wavelength signal detection. Section 2.2 covers the general concepts and definitions about detector principles. Section 2.3 discuses about the terahertz and infrared quantum cascade detectors (QCDs) and at the end of this section two novel dual-wavelengths photodetectors will be presented. In Sect. 2.4, a terahertz quantum well photodetector based on two-photon absorption will be introduced. Section 2.5 gives the brief review on quantum dot long-wavelength photodetectors and also a new quantum dot terahertz photodetector based on the concept of defect will be proposed. EIT-based photodetection and possible structures for long-wavelength photodetector based on physical electromagnetically induced transparency phenomena (EIT) will be introduced in Sect. 2.6.

2.2 Detector Principles

Detecting light means generating an electrical signal (current or voltage) as function of the incident light intensity. The incident photons in a photodetector generate electrons. The number of electrons passing through the device is relevant to the number of photons. This relevance is "the overall quantum efficiency" and expressed as [2]:

$$\eta_{tot} = \frac{N_e}{N_p} \quad (2.1)$$

Since determining the number of electrons and photons is practically almost impossible, measuring the ratio of the output electrical current per incident optical power which is called the responsivity $\Re$ parameter, is preferred:

$$\Re = \frac{I}{P_i} = \frac{\lambda e}{hc} \frac{N_e}{N_p} = \frac{\lambda e}{hc} \eta_{tot} \quad (2.2)$$

where e, h, c and λ are the electron charge, the Planck's constant, the speed of light and the wavelength of the incident light, respectively. Three probabilities are involved in determining the overall quantum efficiency: the absorption probability of an electron passing through the active region that is defined as the following quantum efficiency format:

$$\eta_{abs} = 1 - e^{-\alpha d} \quad (2.3)$$

where α is the absorption coefficient and d the thickness and the escape probability, p_e which is the fraction of electrons excited by absorbing incident photons, contributing to electrical current- and the capture probability, p_c are expressed in the gain format (like the photoconductive gain in QWIP):

$$g = \frac{p_e}{N p_c} \quad (2.4)$$

where N is the number of periods in a multiple period QWIP structure.

2.2.1 Noise Affects

The photodetection process is inevitably affected by noise. A part of photodetector output current is correlated to the intensity of the input light (signal current) but the other part of this current has no correlation with the incoming light. This is defined as "Signal-to-noise-ratio" *S/N*:

$$S/N = \frac{I_{\text{signal}}}{I_{\text{noise}}} \tag{2.5}$$

The power of the input light needed to produce the same signal strength as produced by noise sources is defined as noise equivalent power (NEP):

$$\text{NEP} = \frac{P}{S/N} \quad (I_{\text{noise}} = \Re \cdot \text{NEP}) \tag{2.6}$$

Need to mention that NEP is directly proportional to $\sqrt{A\Delta f}$, which is unrelated to the material properties. Therefore it is more convenient to compare different detectors with the quantity:

$$D^* = \frac{\sqrt{A\Delta f}}{\text{NEP}} \tag{2.7}$$

which is defined as the detectivity. A larger detectivity means a smaller NEP and hence a more sensitive detector. In above equation *A* is the detector's surface and Δf the electrical bandwidth. Inserting Eq. 2.6 gives:

$$D^* = \frac{\Re\sqrt{A\Delta f}}{I_{\text{noise}}} \tag{2.8}$$

The normalized detectivity is expressed in units of "Jones" (cm $\sqrt{\text{Hz}}$/W). The important origins of noise current in a photodetector are: Johnson noise, $1/f$ noise and shot noise due to current flowing through a device. The shot noise is corresponded to dark current noise (without illumination) or photon noise. The $1/f$ noise becomes unimportant in photodetectors if measuring characteristic frequencies are set high enough. In a resistive device, the thermal motion of electrons makes Johnson noise:

$$P_{\text{noise}} = 4k_{\text{B}}T \cdot \Delta f \tag{2.9}$$

For an ohmic device, the current and voltage noise spectral densities are:

$$i^2_{\text{noise},J} = \frac{4k_{\text{B}}T\Delta f}{R}, \quad v^2_{\text{noise},J} = 4k_{\text{B}}TR\Delta f \tag{2.10}$$

where *R* is the device resistance. For Johnson noise limited case, the detectivity is:

$$D_J^*(v) = \Re(v)\sqrt{\frac{R_0 A}{4k_B T}} \tag{2.11}$$

In QWIPs, dark current noise and photon noise are rather important than Johnson noise. In QCDs, dark current is absent so Johnson noise is dominant at high temperatures.

The generation-recombination noise mean square current that dark current I_{dark} makes is described by:

$$i_{\text{noise,dark}}^2 = 4eg_{\text{noise}} I_{\text{dark}} \Delta f \tag{2.12}$$

In the conventional photoconductors, there is almost no difference between noise gain g_{noise} and the photoconductive gain, but for QWIPs they are different. For the current limited case the detectivity can be described as:

$$D_{\text{dark}}^*(v) = \frac{\Re(v)}{\sqrt{u e g_{\text{noise}} J_{\text{dark}}}} \tag{2.13}$$

The background blackbody radiation generates a current, J_{BG}. This current produces noise as well as the dark current. The J_{BG} is given by:

$$J_{\text{BG}} = e \int \eta_{\text{tot}}(\lambda) \frac{d\Phi_{\text{BG}}(\lambda)}{d\lambda} d\lambda = e \int \eta_{\text{tot}}(v) \frac{d\Phi_{\text{BG}}(v)}{dv} dv \tag{2.14}$$

where $\frac{d\Phi_{\text{BG}}(v)}{dv}$ is the background photon flux spectral density. The blackbody emissivity is described as:

$$\begin{aligned} K(v, T_{\text{bb}})dv &= \frac{2hv^3}{c^2} \frac{1}{e^{\frac{hv}{k_B T_{\text{bb}}}} - 1} dv \\ K(\lambda, T_{\text{bb}})d\lambda &= \frac{2hc^2}{\lambda^5} \frac{1}{e^{\frac{hv}{k_B T_{\text{bb}}}}} d\lambda \end{aligned} \tag{2.15}$$

by

$$\begin{aligned} hv \cdot \frac{d\Phi_{\text{BG}}(v)}{dv} &= \pi \sin^2 \frac{\Theta}{2} \cdot K(v, T_{\text{bb}}) \\ \frac{hc}{\lambda} \cdot \frac{d\Phi_{\text{BG}}(\lambda)}{d\lambda} &= \pi \sin^2 \frac{\Theta}{2} \cdot K(\lambda, T_{\text{bb}}) \end{aligned} \tag{2.16}$$

where T_{bb} is the blackbody temperature and Θ is a full cone of view (FOV) angle. Putting J_{dark} and (2.4) in (2.13), we obtain:

$$\begin{aligned} D_{\text{BLIP}}^*(v) &= \frac{\Re(v)}{\sqrt{ue^2 g_{\text{noise}} \int \eta(v') \frac{d\Phi_{\text{BG}}(v')}{dv'} dv'}} \\ &= \frac{\eta(v)}{hv \sqrt{u g_{\text{noise}} \int \eta(v') \frac{d\Phi_{\text{BG}}(v')}{dv'} dv'}} \end{aligned} \tag{2.17}$$

2.2.2 Background Limited IR Performance

Temperature has various effects on the different noise sources. *For* constant $\Re$ temperature increases Johnson noise. Although increasing temperature decreases $\Re$ in semiconductors, Johnson noise likewise increases with temperature. The effect of temperature on the dark current is the same then dark current noise also increases with temperature. The temperature has approximately no effect on photon noise until device temperature below the temperature of the environment.

In photodetectors, there exists a temperature, below which photon noise becomes dominant and above that the other noise sources have dominance. This temperature is indexed by background limited IR performance, T_{BLIP}. Since the different noise sources are independent thus the total detectivity is given by:

$$D^*_{\text{tot}} = \frac{\Re}{\sqrt{\frac{4k_B T}{R_0 A} + ueg_{\text{noise}}\left(J_{\text{dark}} + e \int \eta(v')\frac{d\Phi_{\text{BG}}(v')}{dv'}dv'\right)}} \quad (2.18)$$

2.2.3 Quantum Cascade Detectors

In these detectors, there are two options for a light-induced electron: relaxing to ground state, or tunneling into the next period. Following equation is given for responsivity of this type of photodetectors:

$$R = \frac{\lambda e}{hc}\eta_{abs}\frac{p_e}{N} \quad (2.19)$$

In extracting above equation, the probability of the electron to be captured by the next-period ground state, p_c is assumed unity. The escape probability, p_e is determined by the fraction of light-induced electrons tunneling to the next period.

2.2.3.1 Model of the Electronic Transport Through QCD Structure and Derivation of the Essential Relations

The zero voltage resistance, $R_0 A$ (where R_0 is the resistance of the pixel and A the area of the pixel) is one of the important figures of merit in photovoltaic detectors to characterize the dark current in the absence of optical excitation. It has discussed that due to the well and barrier widths, only interactions between electrons and LO phonons should take into account [3]. On the other hand, due to sufficiently high energy difference between the energy levels, one can omit the effect of interaction between electrons and acoustical phonons [4] and also electron–electron interaction because of subband separation in the cascade structure which is in the range of LO phonon.

The transition rates are evaluated due to the interaction between electrons and optical phonons according to Ferreira and Bastard [5]. Starting from an initial state of wave vector k and energy E in the subband i, the transition rate $S_{ij}^{a,e}(E)$ towards the subband j (in S^{-1}) is obtained through the integration of a matrix element involving a standard electron–optical-phonon Hamiltonian. This integration involves all the possible final states of energy $E \pm \hbar\omega_{LO}$ in the subband j, where $\hbar\omega_{LO}$ is the energy of a LO phonon, the plus or minus sign accounting for absorption or emission of LO phonons, corresponding to superscript a or e, respectively. Transitions rates S_{ij}^a and S_{ji}^e are linked by the following equation [3]:

$$S_{ij}^a(E) = S_{ji}^e(E + \hbar\omega_{LO}) \tag{2.20}$$

The transition rate from an initial state $|i, k_i\rangle$ to all final states $|f, k_f\rangle$ due to longitudinal-optical-phonon emission at $T = 0$ K is equal to [5]:

$$\frac{1}{\tau_i} = \frac{m^* e^2 w_{LO}}{2\hbar^2 \varepsilon_p} \sum_f \int_0^{2\pi} d\theta \frac{I^{ij}(Q)}{Q} \tag{2.21a}$$

$$Q = \left(k_i^2 + k_f^2 - 2k_i k_j \cos\theta\right)^{1/2} \tag{2.21b}$$

where

$$k_f^2 = k_i^2 + \frac{2m^*}{\hbar^2}(\varepsilon_i - \varepsilon_f - \hbar w_{LO}), \tag{2.22}$$

$$\varepsilon_p^{-1} = \varepsilon_\infty^{-1} - \varepsilon_s^{-1} \tag{2.23}$$

where ε_∞ and ε_s are the high-frequency and static relative permittivities of the hetero-layer. $I^{ij}(Q)$ is defined as:

$$I^{ij}(Q) = \int dz \int dz' \chi_i(z) \chi_j(z) e^{-Q|z-z'|} \chi_i(z') \chi_j(z') \tag{2.24}$$

which is equal to δ_{ij} if $Q = 0$ and which decays like Q^{-1} at large Q values. Since the optical-phonon dispersion is neglected in our calculations, the transition rate for phonon emission at nonzero temperature is obtained from (2.21a, 2.21b) by multiplying the zero-temperature result by $(1 + n)$, where n is the thermal population of optical phonons:

$$n = \left[\exp\left[\frac{\hbar w_{LO}}{kT}\right] - 1\right]^{-1} \tag{2.25}$$

The global transition rate G_{ij} between the subband i and subband j is the sum of the two transition rates for absorption of LO phonons $\left(G_{ij}^a\right)$, and emission of LO

2.2 Detector Principles

phonons $\left(G_{ij}^e\right)$. In order to calculate the global transition rates G_{ij}^a and G_{ij}^e, all the initial states of energy E are filled at thermal equilibrium by the Fermi–Dirac occupation factor f. The electronic promotion from a subband i to a higher subband j (i.e., $j > i$) is calculated in the following manner. The integration on all these states is now performed on the subband i:

$$G_{ij}^a = \int_{E_j - \hbar\omega_{LO}}^{+\infty} S_{ij}^a(E) f(E)[1 - f(E + \hbar\omega_{LO})] n_{opt} D(E) dE \qquad (2.26)$$

$$G_{ij}^e = \int_{\varepsilon_j + \hbar\omega_{LO}}^{+\infty} S_{ij}^e(E) f(E)[1 - f(E + \hbar\omega_{LO})] (1 + n_{opt}) D(E) dE \qquad (2.27)$$

$f(E)$ and $f(E + \hbar\omega_{LO})$ are the Fermi–Dirac occupation factors at E and $(E + \hbar\omega_{LO})$, $D(E)$ is the two-dimensional density of state of the subband i and ε_j is the minimum of energy of the subband j. Of course, similar expressions can be written in the case of electronic transfers from a subband i to a lower subband j and n_{opt} is the Bose–Einstein statistic function which accounts for the phonon population.

In order to continue the modeling discussion, we consider the QCD structure introduced by Koeniguer et al. [3] which is presented in Fig. 2.1.

Two consecutive cascades a and b are represented in Fig. 2.2. Figure 2.1 present the most significant transitions where the intracascade and intercascade transitions are separately illustrated in parts (a) and (b), respectively where the solid-line arrows show the main transfer rates between one cascade and the following one and the dashed-line arrows represent other minor transitions. The typical value of transitions rates between two neighboring levels are between a few 10^{20} and 10^{25} m^{-2} s^{-1}, corresponding to transition times between a few ps and a few tens of ps for a temperature of 80 K. These transitions are now limited to a few 10^{18} m^{-2} s^{-1} at the same temperature (and a corresponding transition time greater than 1 µs due also to the low amount of electron promotion to higher subbands for satisfying the Fermi–Dirac distribution). This shows that the electronic mobility is higher inside a cascade than between two consecutive cascades by several orders of magnitude. Finally, the resistance of a QCD is completely determined at 80 K by a few intercascade cross transitions, namely $E_1^A \rightarrow E_5^B$ and $E_1^A \rightarrow E_6^B$. The optimization of a QCD requires decreasing these transitions rates, thus increasing the resistance and decreasing the noise. This is possible by a separation of the wave functions, but at the expanse of a lower optical matrix element, and a lower response. All the challenge of the QCD design consists of mastering this trade off.

The global current density can be evaluated by counting the electronic transitions between the two consecutive cascades. As a consequence, transition rates from cascade A to cascade B (two successive period) are not equal to the

Fig. 2.1 Presentation of a period of a typical QCD structure. **a** Conduction band diagram and wave function associated with each energy level of a period and **b** Principle of a detection [3]

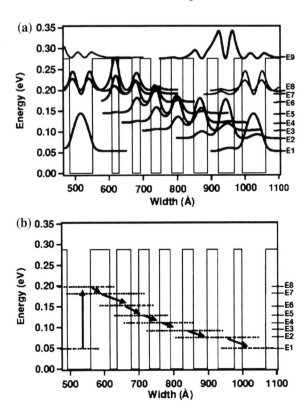

reciprocal transitions rates from cascade B to cascade A. This global current density is given by:

$$J = q \sum_{i \in A} \sum_{j \in B} \left[G_{ij}(V) - G_{ji}(V) \right] \quad (2.28)$$

Let us consider two subband i and j associated, respectively, with cascades A and B. The introduction of two quasi-Fermi levels implies the differentiation of the two Fermi occupation factors in Eqs. 2.25 and 2.26. G_{il}^a and G_{ji}^e can be evaluated by:

$$G_{ij}^a(V) = \int_{\varepsilon_j - \hbar\omega_{LO}}^{+\infty} S_{ij}^a(E) f_A(E) [1 - f_B(E + \hbar\omega_{LO})] n_{opt} D(E) dE \quad (2.29)$$

$$G_{ij}^e(V) = \int_{\varepsilon_j}^{+\infty} S_{ij}^e(E) f_B(E) [1 - f_A(E + \hbar\omega_{LO})] (1 + n_{opt}) D(E) dE \quad (2.30)$$

where f_A and f_B are the Fermi–Dirac occupation factors associated with quasi-Fermi level E_F^A and E_F^B. Considering Eq. $(S_{ij}^a(E) = S_{ji}^e(E + \hbar\omega_{LO}))$ the difference is then equal to:

2.2 Detector Principles

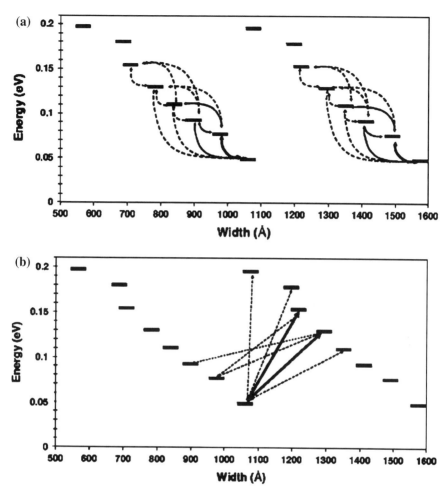

Fig. 2.2 Major transition rates of two consecutive cascades of the real device at 80 K. **a** The main transition rates inside each cascade (only the transition rates greater than 10^{20} m^{-2} s^{-1} are represented) and **b** the main transitions between the cascades: *solid lines* concern the major transition (greater than 4×10^{18} m^{-2} s^{-1}), whereas *dashed lines* represents the other main transitions (greater than 1×10^{18} m^{-2} s^{-1} and lower than the major transitions) [3]

$$G_{ij}^a(V) - G_{ij}^e(V) = \int_{\varepsilon_j - \hbar\omega_{LO}}^{+\infty} S_{ij}^a(E) a(E) [1 - \gamma(E)] dE \qquad (2.31)$$

where

$$\begin{aligned} a(E) &= n_{opt} f(E)[1 - f_B(E + \hbar\omega_{LO})] D(E), \\ \gamma(E) &= \frac{f_B(E + \hbar\omega_{LO})[1 - f_A(E)](1 + n_{opt})}{f_A(E)[1 - f_B(E + \hbar\omega_{LO})] n_{opt}} \end{aligned} \qquad (2.32)$$

Without any applied voltage, the term $a(E)$ is equal to $a^{eq}(E)$ given by:

$$a^{eq}(E) = n_{opt}f(E)[1 - f(E + \hbar\omega_{LO})]D(E) \quad (2.33)$$

Expressing the Fermi–Dirac functions, $\gamma(E)$ is simplified into:

$$\gamma(E) = \exp\left(\frac{E_F^B - E_F^A}{k_B T}\right) = \exp\left(\frac{-qV}{k_B T}\right) \quad (2.34)$$

where T is the temperature of the sample and k_B the Boltzman constant.

We recall that in the context of infrared photovoltaic detection, applied biases are very small. The Johnson noise is related to the resistance at 0 V: R_0. In this calculation, it is then justified to linearize:

$$1 - \gamma(E) \approx \frac{q}{k_B T}V \quad (2.35)$$

This leads directly to the linear $I(V)$ behavior of the structure at low bias, through the multiplication by the constant a^{eq} (calculated with no applied voltage). For little variations of the voltage, the difference can be approximated by the following equation:

$$G_{ij}^a(V) - G_{ji}^e(V) \approx \int_{\varepsilon_j - \hbar\omega_{LO}}^{+\infty} S_{ij}^a(E)\alpha^{eq}\frac{qV}{k_b T}dE$$

$$\approx G_{ij}^a(V = 0V)\frac{qV}{k_B T} \quad (2.36)$$

Finally, global current density is so evaluated by the formula:

$$J = q\sum_{i \in A}\sum_{j \in B} G_{ij}\frac{qV}{k_B T} \quad (2.37)$$

where the term G_{ij} is defined by the sum of G_{ij}^a and G_{ji}^e calculated without any applied voltage. R_0A can be finally deduced from the last equation:

$$R_0A = \frac{k_B T}{q^2 \sum_{i \in A}\sum_{j \in B} G_{ij}} \quad (2.38)$$

The linear absorption coefficient, $\alpha(\omega)$ for the intersubband transitions can be clearly calculated as follows [6]:

$$\alpha(\omega) = \frac{\omega\mu c e^2}{n_r}|M_{fi}|^2\frac{m^* k_B T}{L_{eff}\pi\hbar^2}Ln\left\{\frac{1 + \exp[(E_F - E_i)/k_B T]}{1 + \exp[(E_F - E_f)/k_B T]}\right\}$$

$$\times \frac{\hbar/\tau_{in}}{(E_f - E_i - \hbar\omega)^2 + (\hbar/\tau_{in})^2} \quad (2.39)$$

where $E_{\text{fi}} = (= \Delta E) = E_f - E_i$, E_i and E_f denote the energy levels for initial and final states, respectively. Also, $M_{\text{fi}}, \mu, L_{\text{eff}}, n_r$ and τ_{in} are the dipole matrix element between initial and final states, the permeability, the spatial extent of electrons in subbands, the refractive index and the intersubband relaxation time, respectively.

For all types of quantum detectors [7] the responsivity is given by:

$$R = \eta \frac{\lambda e}{hc} \qquad (2.40)$$

where η is the collection efficiency involving the photon absorption probability and the photoexcited electron transfer probability and λ the wavelength. One can assume the parameter η to be the external quantum efficiency of the device which corresponds to the ratio between the number N_c (electrons stored in the capacitor of the readout circuit) and the number N_{ph} (incident photons during the same time): $\eta = N_c/N_{\text{ph}}$. Introducing the number N_e of excited electrons in the whole QCD structure (that is in the N periods), the external quantum efficiency can be written as the product of two ratios: $\beta = N_c/N_e$ and $\alpha = N_e/N_{\text{ph}}$.

$$\eta = \frac{N_c N_e}{N_e N_{\text{ph}}} \qquad (2.41)$$

β gives the proportion of the excited electrons which are finally stored in the external capacitor with respect to the quantity of infrared-excited electrons: β is the so-called photoconductive gain composed of $N = 40$ periods, 40 excited electrons are needed to generate one electron only in the readout circuit. This defines a maximum photoconductive gain of $1/40 = 2.5\%$. α is the ratio between the number of excited electrons in the entire device, which corresponds to the total number of absorbed photons in the entire structure, and the total number of incident photons: α is the absorption coefficient, also called quantum efficiency in the community of mid-IR detectors.

From responsivity and resistivity values, we can deduce the Johnson noise limited detectivity given by:

$$D^* = R(\lambda)\sqrt{\frac{R_0 A}{4kT}} \qquad (2.42)$$

where $R(\lambda)$ is the peak responsivity, R_0 is the device resistance at null bias, A is the mesa area and T the temperature of the sample.

2.2.4 Effects of Number of Periods and Doping Density on the Detector Parameters

Quantum efficiency in a multiple period QWIP structure is expressed as [2]:

$$\eta_{abs} = 1 - e^{-N\alpha_{2D}\frac{\sin^2\Theta}{\cos\Theta}} \tag{2.43}$$

where N is the number of periods. If we assume, that η_{abs} is much smaller than 1, the e^{-x} can be approximated by $1-x$. Therefore we have:

$$\eta_{abs} = N\alpha_{2D}\frac{\sin^2\Theta}{\cos\Theta} \propto N \tag{2.44}$$

Thus responsivity will be:

$$\Re \propto \eta_{abs}\frac{1}{N} \propto N \times \frac{1}{N} = \text{constant} \tag{2.45}$$

As seen in above equation, the responsivity is independent of the number of periods, N. However the noise gain changes with N as:

$$g_{noise} \propto 1/N \tag{2.46}$$

Therefore the background photon limited detectivity D^*_{BLIP} varies as:

$$D^*_{BLIP} = f(\Re, g_{noise}) \propto \sqrt{N} \tag{2.47}$$

In a QCD, responsivity is again constant but the device resistance R_0 will increase with N and hereby the Johnson noise limited detectivity, D^*_J will change as:

$$D^*_J \propto \sqrt{N} \tag{2.48}$$

It should be mentioned that D^*_{BLIP} the same behavior. Stacking N periods of a detector structure together means putting N current (noise) sources in series leads reduces the total noise by the $\sqrt{N}$. Increasing the number of periods does not change the BLIP temperature T_{BLIP}, but increases the total detectivity D^*_{tot} by $\sqrt{N}$.

In a similar way it is shown for small absorption, the absorption and then responsivity rises linearly in doping density, n_s as:

$$\Re \propto n_s \tag{2.49}$$

The background photon limited detectivity D^*_{BLIP} changes as:

$$D^*_{BLIP} \propto \sqrt{n_s} \tag{2.50}$$

Although in QWIP, the dependence of dark current on doping density is explicit, but since the transport system in QCD contains the large number of electronic states so there is no a simple and clear expression for the device conductance $G = 1/R_0$. Howbeit the population of the different states is determined by doping, the overlap of the states wavefunctions mainly determines the detector overall performance. Koeniguer et al. [3] have reported a model for a QCD structure, to simulate the n_s dependence of R_0A and therefore D^*_J.

2.2.5 Quantum Dot Infrared Photodetectors

In what follows, the calculation of detector parameters of this photodetectors is described. The peak responsivity in quantum dot infrared photodetectors (QDIP) is given by:

$$R_p^o = \left(\frac{e}{h\nu}\right)\eta_\alpha p_e g, \qquad (2.51)$$

where $h\nu$ and g are photon energy and optical gain which is given by Beck [8] in terms of the fill factor F, capture probability p_c and number of QD layers N, respectively and given as:

$$g = \frac{1 - \frac{p_c}{2}}{FNp_c}, \qquad (2.52)$$

Also, the absorption quantum efficiency η_α can be expressed in term of high-field domain length ℓ as follows:

$$\eta_\alpha = \left(1 - e^{-2\alpha\ell}\right) \qquad (2.53)$$

Finally escape probability is expressed in the following relation:

$$p_e = \left[1 + \frac{\tau_t(v)}{\tau_r}\right]^{-1}, \qquad (2.54)$$

where τ_r and τ_t are recapture life time and tunneling time which is given by Levine [9] in terms of the dot size L_D, phase velocity of an electron near the first excited state v and transmission coefficient $T(v)$, respectively as:

$$\tau_t = \frac{2L_D}{vT(v)} \qquad (2.55)$$

The dark current in a QDIP, as a function of applied bias V is given by [10]:

$$I_{\text{dark}}(V) = ev_d(V)n_{\text{total}}(V)A \qquad (2.56)$$

where $v_d(V)$ is the average electron drift velocity in the barrier material, n_{total} is the concentration of electrons excited out of the quantum dots by the thermionic emission and tunneling, and A is the detector area. Here:

$$v_d(V) = \frac{\mu F(V)}{\sqrt{1 + \frac{\mu F(V)}{v_s}}}, \qquad (2.57)$$

$$n_{\text{total}}(V) = n_{3D}(V) + n_{\text{dot}}(V), \qquad (2.58)$$

where the three-dimensional electron density $n_{3D}(V)$ can be estimated by Liu [11] in terms of the barrier effective mass m_b and thermal activation energy

$E_a (= hc/\lambda_{\text{cut-off}})$, which equals the energy difference between the top of the barrier and the Fermi level in the dot as:

$$n_{3D} \simeq 2\left(\frac{m_b k_B T}{2\pi\hbar^2}\right) e^{\frac{E_a}{k_B T}} \quad (2.59)$$

Also, the zero-dimensional electron density which can tunnels through the barrier $n_{\text{dot}}(V)$ was calculated in [12] and illustrated in Eq. 2.60.

$$n_{\text{dot}}(V) = \int_0^{V_{02}} N_{\text{dot}}(E) f(E) T(E,V) dE \quad (2.60)$$

In the calculation of Eq. 2.59, we assumed that the transmission coefficient $T(v)$ is unit for $E > V_{02}$. In Eq. 2.60, N_{dot} is density of states of the quantum dots. The total noise for a photodetector can be expressed as sum of generation-recombination (shot) and the Johnson (thermal) noise terms. The Johnson noise is not included in the calculation, since it is generally much less than the shot noise [13–15]. The photoconductor current shot–noise can now be obtained using the optical gain g (valid for small quantum dot capture probability, i.e., $p_c \cong 1$):

$$i_n = \sqrt{4eI_{\text{dark}} g \Delta f}, \quad (2.61)$$

where Δf is the bandwidth. Now the specific detectivity D^* of QDIP at different temperature and applied biases is obtained from the peak responsivity and current shot-noise [12] as follows:

$$D^* = R_p^o \sqrt{A_{\text{eff}} \Delta f} / i_n, \quad (2.62)$$

where A_{eff} is the absorption effective cross-section area. Finally one can calculate the noise-equivalent temperature difference NEΔT as [14]:

$$\text{NE}\Delta\text{T} = i_n k T_B^2 \lambda \bigg/ \left[hc A_{\text{eff}} \sin^2(\Omega/2) \int_{\lambda_1}^{\lambda_2} R(\lambda) W(\lambda) d\lambda \right] \quad (2.63)$$

where Ω is optical field of view, T_B is the black body temperature, λ_1 and λ_2 are the integration limits that extend over the responsivity spectrum and $W(\lambda)$ is the blackbody spectral density which is obtained as:

$$W(\lambda) = (2\pi c^2 h/\lambda^5)(\exp(hc/\lambda k T_B) - 1) \quad (2.64)$$

2.3 Terahertz and Infrared Quantum Cascade Detectors

Quantum cascade structures are relatively new structure in the field of QWIP which are introduced to reduce the dark current owed to the bias-free operation (photovoltaic operation mode). Thus the capacitor loading problem in the focal plane arrays due to the dark current and the dark current noise problem are minimized.

Quantum well infrared photodetectors (QWIPs) have been developed in the recent years from the fundamental-physics point of view towards fabrication of large area focal plane arrays [1, 16, 17]. In principle, the QWIPs operate in two electronic transport modes: photoconductive and photovoltaic. The first operation mode requires external bias voltage to conduce the electrical current where the excited electrons tunnel out of the quantum wells in the presence of the electric field and are gathered as a measure of incident light. The limiting factor in this mode of operation is the presence of dark current which bound the device performance to low temperatures. The typical applied electric fields are about several 10 kV/cm which are the direct result of a trade-off between high quantum efficiency (the capture probability of photo-excited carriers) and low noise. The problem of dark current is a motivation for developing low dark current and appropriate quantum efficiency (the second operation mode).

Beside the mentioned problems, the carrier transport mechanism in photoconductive mode involves both 2D and 3D electron states in well and continuum, respectively, between which the modeling of carrier transition has its own theoretical complications [18, 19]. The diffusion of electrons from 3D to 2D states (and vice versa) is a particularly difficult theoretical problem. That is why most models use adjustable parameters such as the capture time [20] or the capture probability, affecting the photoconductive gain [21].

The second types of photodetectors, i.e., photovoltaic photodetectors, do not require external bias voltage and thus suffer very little or even no dark current. Therefore, there is no generation-recombination noise in the dark condition (and therefore the capacitors in the focal plane arrays are not filled by dark current).

A conceptual schematic of a photovoltaic photodetector structure fabricated by Kastalasky et al. [22] is presented in Fig. 2.3a. The operation principle of this device is based on electron capture from a miniband which terminates with a blocking barrier. Since the electron current is blocked by the blocking barrier in the lower miniband, only electrons of higher energies (in the upper miniband) have enough energy to pass through the barrier and thus the photo-induced current is created without need to bias voltage. Goossen et al. [23] have introduced a different structure (Fig. 2.3b) in which the device capacitance depends to the carriers flowing through the depletion layer. A quantum well-based structure with asymmetrically doped double barriers composed of GaAs/AlAs/AlGaAs material compositions is schematically plotted in Fig. 2.3c [24]. The photo-excited electrons from the lower to the upper miniband tunnel out of the quantum well with a finite probability through the AlAs barrier. There are two probabilities whether the excited electrons transport to left or right side. The spatial field in the AlGaAs layers causes an electric current to the right side and most of carriers are relaxed in their path to the quantum well place in the right-hand side of $Al_{0.3}Ga_{0.7}As$ layer. The general concept of this transport mechanism is illustrated in Fig. 2.3d. Figure 2.3e also presents a similar structure to the structure in Fig. 2.3c but with a reduced barrier height. The photovoltaic operation of this structure is driven by a

Fig. 2.3 Transport mechanism of photovoltaic infrared detectors involving intersubband transitions: **a** superlattice with blocking barrier, **b** single quantum well with surface depletion layer, **c** asymmetrically doped double-barrier quantum well, **d** general subband configuration for photovoltaic detection, and **e** modulation-doped single-barrier quantum well [27]

Fig. 2.4 a Schematics of the four-zone approach for photovoltaic Intersubband photodetection; potential distribution (*1* emission zone, *2* drift zone, *3* capture zone, and *4* tunneling zone) and basic operation (*arrows*). **b** Bandedge distribution of a four-zone QWIP in an electric field and considerations for optimization [27]

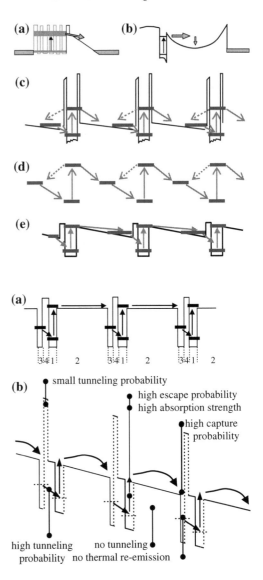

tunneling barrier and a modulation doped induced internal electric field [25, 26]. The detectivity parameter of this structure is three times lower than the reported values for photovoltaic infrared photodetector operating at the same wavelength (10.5 μm). Low detectivity of this structure may be relates to the non-optimized internal electric field.

A new four-zone structure for QWIP has been proposed by Schneider et al. (Fig. 2.4) where each period of the active region consists of four independent zones: (1) excitation zone, (2) drift zone, (3) capture zone and (4) tunneling zone [28, 29]. The photo-excited carries in the excitation zone are transported in the quasi-continuum band edge of the drift region. Then the carriers are relaxed to the capture zone and fill the quantum well by tunneling. This contributes to a photocurrent without any external bias voltage.

Quantum cascade structure are of the well-known structures for effective transport mechanism. A photovoltaic intersubband quantum cascade photodetector (QCD) has been introduced in Ref. [30] which operates without any bias voltage and hence suffers no dark current. The optical and electrical transport specifications of these structures confirm the suitability for small pixel-large array applications. The main design purpose of the QCD is creation of a bias-free photo-induced electron displacement through the cascaded quantum states. Thus, a comprehensive study is required to get familiar with the carrier transport process in these structures. Generally quantum cascade devices have been described with kinetic model band making use the Monte Carlo simulation [31].

2.3.1 Dual Color Mid-Infrared Quantum Cascade Photodetector in a Coupled Quantum Well Structure [32]

Beside the positive and negative points of the mentioned structures, simultaneous detections of two or more wavelength can be an interesting feature. Several applications are predicted for multi-wavelength detection such as reducing the number of false positives through detection of infrared radiations of an object [33]. Also, dual-wavelength imaging in terahertz and mid-IR region is another interesting feature of multi-wavelength detection [34].

In this section, we introduce a QCD structure which is capable of detecting two simultaneous wavelengths (9.94 and 5.88 μm).

Figure 2.5 shows the respective QCD structure. This structure is composed of 40 periods of coupled AlGaAs/GaAs QWs. The first QW is n-doped in order to populate its first level of energy E1 in the conduction band with electrons. The second well is as the capturing well, attached to main active well. The absorption of a photon at energy $h\nu = E_6 - E_1$ and $h\nu = E_7 - E_1$ transfers an electron from ground state towards sixth and seventh excited states. The first transition will be transported after coupling in the second well of the transport path constructed by superlattice structure. The second transition will be captured to the second well by release of a LO phonon and will be transported through the same transport path constructed by superlattice structure.

Table 2.1 provides the global transition rates G_{1j} for novel QCD structure, the number of transitions per second and per square meter from the fundamental level 1 to the level j, with $j = \{2, \ldots, 7\}$, in the neighboring cascade. For temperatures 80, 120, 250 K, the results have been obtained.

Figure 2.6 presents a typical example chosen to illustrate the quality of the modeling: the experimental R_0A of the device as a function of $1000/T$, where T is the temperature of the sample.

Fig. 2.5 The schematic of dual-wavelength QCD structure (size of considered layers from left to right in Å are: 22,**65**,22,**40**,5,10,10,10,5,**36**, 22,**42**,22,**50**,22,**65**,22. The under lined layer, denotes doped well to 5×10^{-5} cm^{-2} and bold layers denote quantum wells) [32]

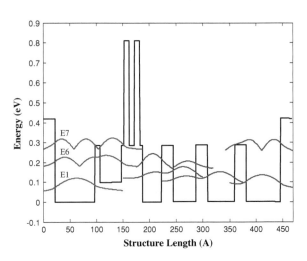

Table 2.1 Values of some transition rates in relation to the main electronic transition represented on Fig. 2.5

$G_{1,j}$ (m^{-2} s^{-1})	80 K	120 K	250 K
1 → 2	5.7932e + 015	3.8970e + 017	3.5614e + 019
1 → 3	3.8508e + 017	2.5900e + 019	2.3630e + 021
1 → 4	4.3563e + 018	2.9307e + 020	2.6815e + 022
1 → 5	1.3963e + 021	9.3969e + 022	8.6427e + 024
1 → 6	1.4009e + 024	9.4284e + 025	8.6775e + 027
1 → 7	2.2470e + 024	1.5110e + 026	1.3733e + 028

Fig. 2.6 R_0A (product of the resistance at 0 V by the area of the pixel) as a function of $1,000/T$ [32]

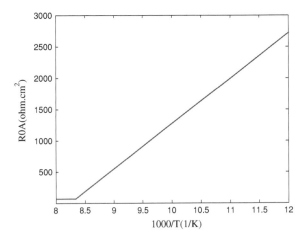

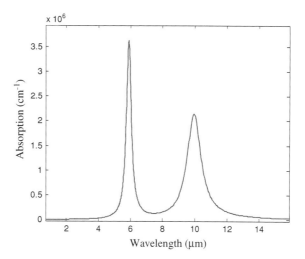

Fig. 2.7 Absorption spectrum at 120 K temperature [32]

Figure 2.7 shows the absorption spectra performed at 120 K temperature. The two peaks can be easily linked to an intersubband transition between two energy levels and the different transitions are identified on this figure: the peak at 9.94 μm and other at 5.88 μm corresponds to the transition from level E1 (ground state) to levels E6 and E7 (sixth and seventh excited states), respectively.

Figure 2.8 illustrates the responsivity spectra calculated at 120 K for two wavelengths. We found two responsivities: $R_{peak,1} = 67.5$ mA/W, $R_{peak,2} = 118.5$ mA/W, at $\lambda = 6.82$ μm and at $\lambda = 12.35$ μm, respectively.

The detectivity spectrum at 120 K for two wavelengths is shown in Fig. 2.9. Taking $R_0A = 70\,\Omega\,\text{cm}^2$ we obtained at 120 K: $D^*(\lambda = 12.35) = 1.2 \times 10^8$ J, $D^*(\lambda = 6.85) = 6.89 \times 10^7$ J.

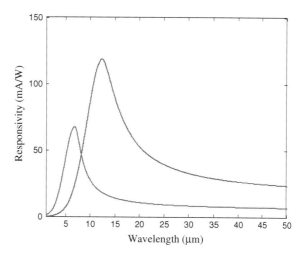

Fig. 2.8 Responsivity spectrum at 120 K for two wavelengths [32]

Fig. 2.9 Detectivity spectrum at 120 K for two wavelengths [32]

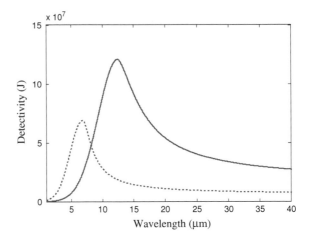

Fig. 2.10 Detectivity spectrum at 80 K for two wavelengths [32]

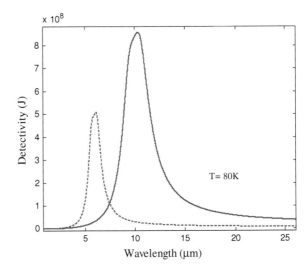

For two different temperatures (80, 250 K) the detectivity spectrum for two wavelengths are represented in Figs. 2.10 and 2.11, respectively.

2.3.2 A Dual-Color IR Quantum Cascade Photodetector with Two Output Electrical Signal

Recently, there has been a great deal of interests in the fabrication of multicolor detectors to enhance the performance of detection, in particular, for discrimination

2.3 Terahertz and Infrared Quantum Cascade Detectors

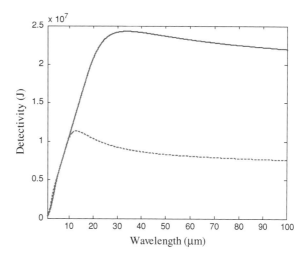

Fig. 2.11 Detectivity spectrum at 250 K for two wavelengths

of objects and imaging under varying atmospheric conditions [35]. As mentioned before the different wavelength regions are associated with different practical applications, detecting an object's infrared emission at multiple wavelengths can be used to reduce the number of false positives. The possibility to have dual wavelength operation is also very attractive in the THz and mid-IR regions for applications like dual wavelength imaging [34]. In order to get response in several wavelength ranges in QWIPs, there are several procedures. One approach is to build several stacks of square quantum wells with different peak response wavelengths [36–43]. The response due to different wavelengths can be achieved either by contacting each stack separately in a constant bias or by controlling the bias across the stacks to sequentially activate different stack. The other approach is to use asymmetric or coupled quantum well structures where the transitions from the ground state to several excited states are allowed [44–46]. The advantage of the latter approach is that it requires only one set of quantum wells which makes the fabrication relatively simpler. In this case, the photoexcited carriers are extracted by controlling the bias across the device. In spite of all achieved developments in multi-color detection, there are several important deficiencies such as bias-dependant wavelength detection, difficulties in simultaneous detection of wavelengths and identical output current path associated with all detected wavelengths.

In this section, we introduce a detector structure based on quantum cascades which is able to detect two different wavelengths simultaneously through two independent output current paths. The two under consideration structures consists of own left and right paths that each path can detect specified wavelength with associated active region. In first structure, the N consecutive periods in each path was studied however, for second structure, we have considered only one period with single main active region for both left and right parts.

The first proposed structure for DC-QCD consists of two separated N-period GaAs/AlGaAs heterostructures containing specified active regions and subsequent

Fig. 2.12 **a** Conduction band profile and associated wave functions for first structure in one period and **b** 3D schematic of the QCD device structure

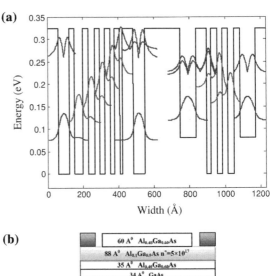

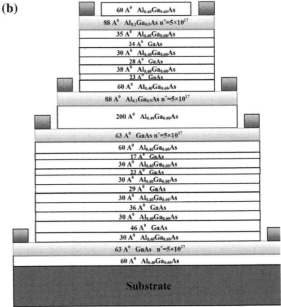

transport ladders in each period to detect wavelengths in completely independent output current paths. A 200 Å $Al_{0.40}Ga_{0.60}As$ layer is used as spacer layer between two sections. The energy levels and associated wave functions for considered structure as well as 3D view of the QCD structure are illustrated in Fig. 2.12 in one period of the device. All QWs and barriers are made of GaAs and $Al_{0.40}Ga_{0.60}As$, respectively (except for active region of right path which is composed of $Al_{0.1}Ga_{0.9}As$). The layer widths in one period from left to right (active region and ladder) in the left path starting with active well layer are: 63/60/17/30/23/30/29/30/36/30/46/30 and 63 Å, respectively.

2.3 Terahertz and Infrared Quantum Cascade Detectors

The right path (in the right-hand side of thick $Al_{0.40}Ga_{0.60}As$ layer) consists of N successive periods with layer widths of 88/60/23/38/28/30/34/35/88 Å, respectively in one period starting with the active well. The peak detection energies of the DC-QCD, which corresponds to the intersubband transition energies between ground and first excited state in each active region was calculated to be 185 meV (6.74 μm) for left path and 107 meV (11.70 μm) for right path.

The second proposed structure for DC-QCD consists of only one active region (stepped quantum well) and with two transport ladders that extract the photo-excited carriers through two side of active well. Figure 2.13 represents the conduction band profile, energy levels and the wave functions associated with each energy level for considered structure.

Barriers, QWs and two steps are made in $Al_{0.40}Ga_{0.60}As$, GaAs and $Al_{0.22}Ga_{0.78}As$, respectively. The layer sequence in Å starting from the first quantum well from left to right is as follows: 63/36/46/32/36/32/29/36/21/33/18/60/42/52/30/60/25/30/30/30/38/30/48/30/64. The intersubband transition energies between ground-first excited state and ground-second excited state in active region were calculated to be 140 meV (8.88 μm) and 223.7 meV (5.56 μm), respectively. The advantage and disadvantage of such structure with only one period are discussed at the end of section. Figures 2.14 and 2.15 presents the resistivity at 0 V (R_0A) as a function of $1000/T$ for right and left path of two proposed structure, respectively.

The short width for right paths in two structures leads to high intercascade transition rates, so the values of R_0A are low for these two paths in comparison with left parts with high values of R_0A. Figure 2.16 presents the linear absorption coefficient, $\alpha(w)$ for the intersubband transitions of considered structures.

Figure 2.17 depicts the peak responsivity for two different structures of proposed DC-QCD.

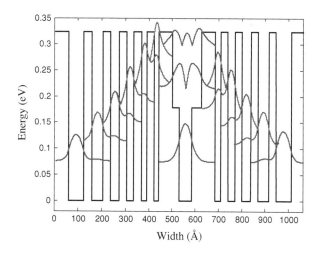

Fig. 2.13 Conduction band profile and wavefunctions for second structure

Fig. 2.14 R_0A for **a** Left path and **b** right path of the first structure (T is the sample temperature)

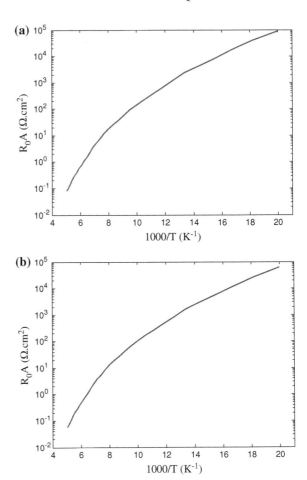

Finally, from responsivity and resistivity, it can deduce the Johnson noise limited detectivity. Figure 2.18 presents detectivity spectrum for one period of two considered structures in $T = 80$ K.

The extracted R_0A figures for right and left paths of first and second structures are considered for only one period but the whole resistance of the device is directly proportional to N ($R_0A_{Total} = N \times R_0A_{one\text{-}period}$) and decrease exponentially with doping concentration. So, for second structure, the obtained R_0A is N times smaller than the device resistance for the first structure, resulting in lower induced potential. On the other hand, the detectivity parameter is relative to N; it means that the detectivity for the first structure is $\sqrt{N}$ times larger than for the second structure. Since the main limitation of quantum cascade detectors is a low response with respect to photoconductive detectors, the main challenge is to design a structure with higher response keeping the same dark current level. This goal is not achievable by increasing the doping concentration since the resistance of the

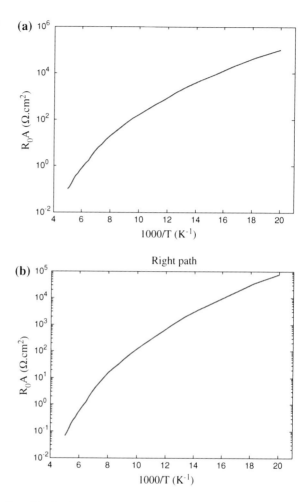

Fig. 2.15 R_0A for **a** left path and **b** right path of second structure

device decreases dramatically with doping concentration. Another approach is the reduction of the number of periods. In this way, an improvement is expected and high responsivity values are achievable for such structures with small number of periods. So, in quantum cascade based detectors, there is a tradeoff between detector parameters meaning that good structure design leads to proper performance of the devices.

2.4 Terahertz Quantum Well Photodetector Based on Two-Photon Absorption [47, 48]

Nonlinear coefficient for second-order susceptibility resulting from intersubband transitions in QWs increases with three orders of magnitude in comparison with host material GaAs [49]. In particular, for two-photon intersubband transitions,

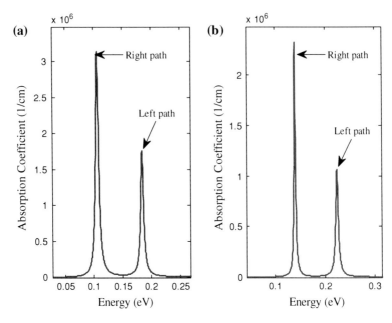

Fig. 2.16 Absorption coefficient as a function of energy **a** for first structure and **b** for second structure ($T = 120$ K)

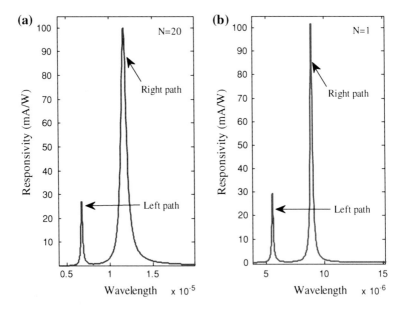

Fig. 2.17 Responsivity spectrum as a function of wavelength **a** for first structure and **b** for second structure ($T = 120$ K)

2.4 Terahertz Quantum Well Photodetector Based on Two-Photon Absorption

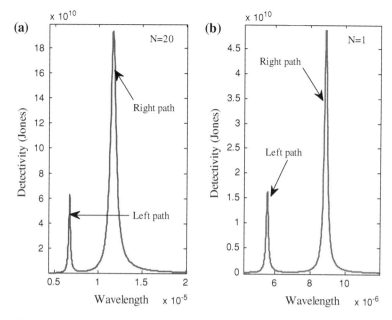

Fig. 2.18 Detectivity spectrum as a function of wavelength **a** for first structure and **b** for second structure ($T = 80$ K)

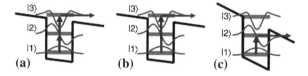

Fig. 2.19 Band diagrams for **a** quadratic detection associated with transitions $|1\rangle \rightarrow |2\rangle$ and $|2\rangle \rightarrow |3\rangle$, **b** linear detection by transition $|1\rangle \rightarrow |3\rangle$, and **c** linear detection involving $|1\rangle \rightarrow |2\rangle$ assisted by tunneling [47]

nonlinearities are six orders of magnitude stronger than the bulk GaAs [50, 51]. The QWIP which is exploited for this, designed for wavelengths at 8–12 μm with two bound subbands $|1\rangle, |2\rangle$, and one equidistant continuum resonance $|3\rangle$ where an electron is excited into the continuum by two infrared photons and hence produced photocurrent is a quadratic function of the incident power (Fig. 2.19a) [50]. This nonlinear behavior introduces a new quadratic detector device in contrast with the linear power dependence QWIPs [52]. Excellent sensitivity for quadratic detection has been achieved, with nonlinear optical signals appearing at power densities as low as 0.1 W/cm² [50]. Since the parasitic time constants arising from the device capacitances or resistances are not influencing the nonlinear process, temporal resolution of two-photon QWIPs is only restricted by the sub-ps intersubband and phase relaxation times.

Lately too much attention has been paid for nonlinear THz detectors because of their applications in femtosecond lasers based THz technology and room temperature based nonlinear detection [53–55].

The structure of two-photon THz-QWIP is made of an active region (consisting 20 GaAs QWs of 18 nm width with central Si-doping of 1×10^{17} cm^{-3} and 70 nm width $Al_{0.05}Ga_{0.95}As$ barriers) sandwiched between two n-type GaAs contact layers with 400 and 700 nm thicknesses. As usual the coupling of the THz radiation into the active region was accomplished using 45° facets [52]. Computations for the chosen structural parameters reveal an operation wavelength slightly less than 50 μm [47].

By using the linear photocurrent spectroscopy which is carried out by a FTIR spectrometer at low temperature and various bias voltages, spectral properties were analyzed (Fig. 2.20). Considering the linear excitation from the ground state into the continuum for the 0.1 V bias voltage the photocurrent begins in at photon energies around 380 cm^{-1} (Fig. 2.19b: the parity forbidden selection rule is not satisfied due to the applied electric field and the state $|3\rangle$ being in the continuum) and eventually this leads to an increase of photocurrent around the cutoff energy. At the 0.5 V bias voltage a significantly steeper increase in the cutoff energy (due to the odd parity of the applied electric field) along with moves towards the lower energies is observed. Because of two-photon excitation being negligible at the low intensities as well as the state $|2\rangle$ being completely confined in the QW at this field, no signal is generated at the energy of the $|1\rangle \rightarrow |2\rangle$ transition. However at 0.94 and 1.5 V bias voltages, the barrier leakage at the energy of $|2\rangle$ (due to Fowler–Nordheim tunneling into the continuum) allows that carriers can escape from the second state of QW with finite probability (Fig. 2.19c) thus induces a pronounced

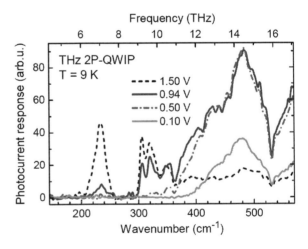

Fig. 2.20 Linear photocurrent spectra of two-photon THz-QWIP at different bias voltages [48]

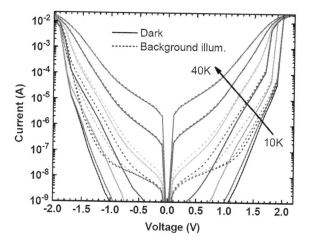

Fig. 2.21 Current–voltage curves of THz two-photon QWIP at temperatures from 10 to 40 K with and without background illumination [48]

peak at 235 cm^{-1} (42.5 μm). The Reststrahlen band (due to strong optical-phonon induced absorption in the GaAs substrate) suppresses the response towards higher energies. A significant photoresponse between the Reststrahlen band and the $|1\rangle \rightarrow |3\rangle$ transition energy is occurred above 300 cm^{-1}. It is worth mentioning that there are a number of narrow dips in the spectral shape above the Reststrahlen band which are caused by two-phonon absorption, typical for THz QWIPs, whereas a Fano lineshape is observed if the active region is involved in the two-phonon transition [56].

Based on the investigations of I–V curves at varying temperatures for two-photon THz-QWIP (Fig. 2.21), the detector is still BLIP (background limited performance) at 25 K if the bias is less than 1 V.

Quadratic detection is being investigated through the measuring of the photocurrent as a function of incident power where a free-electron laser (FEL) excites the transition at a wavelength of 42 μm. The two-photon THz-QWIP is biased at 500 mV and below in order to suppress the tunneling-induced linear photocurrent contribution. The final state of the two-photon transition is aligned with the continuum. Figure 2.22 shows the photocurrent I versus the incident power P of the FEL where the photocurrent increases quadratically with the incident power, followed by saturation behavior. The intensity dependence is described by the equation [57]:

$$I = SP^2 W\left(\frac{I_{Sat}}{SP^2}\right) \quad (2.65)$$

where S represents the quadratic contribution to the photocurrent, I_{Sat} the saturation current, and W Lambert's W-function. This is an approximation where the photoconductive gain depends linearly and the dark current exponentially on the local electric field inside the active region.

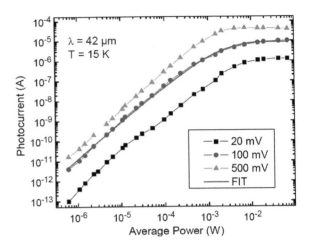

Fig. 2.22 Measured intensity dependence of the photocurrent at 20, 100, and 500 mV bias, including a fit for 100 mV [48]

2.5 Quantum Dots THZ-IR Photodetector

2.5.1 An Overview of Quantum Dot

The introduction of quantum wells in the early 1970s was a turning point in the direction of research on electronic structures [58]. A quantum well is a very thin layer of a semiconductor sandwiched between two layers of another semiconductor with wider energy gaps. The motion of electrons in a quantum-well structure is bound in two directions if the thickness of the quantum well layer is of the order of the de Broglie wavelength [15].

In the 1980s the interest of researchers shifted toward structures with further reduced dimensionality: one-dimensional confinement (quantum wires) [59] and zero-dimensional confinement (quantum dots). Localization of carriers in all three dimensions breaks down the classical band structure of a continuous dispersion of energy as a function of momentum. Unlike quantum wells and quantum wires, the energy level structure of quantum dots is discrete.

To demonstrate useful QD-based devices several requirements need to be fulfilled [60]:

1. Small QD size and sufficiently deep localizing potential for observation of zero-dimensional confinement.
2. High density of QDs and a high filling factor.
3. Low dislocation density material.

The QD size should not exceed a lower and an upper size limit dictated by the population of energy levels. The minimum QD size is the one that ensures existence of at least one energy level of an electron or a hole or both. For instance, for the InAs/AlGaAs system, the critical diameter of the QDs is ~ 3–5 nm, whereas this value is about 1 nm for the GaN/AlN system [15].

2.5 Quantum Dots THZ-IR Photodetector

Several methods have been used for fabrication of QD structures. Depending on the thicknesses associated with the potential confinement, the energy of the intersublevel transitions can exhibit resonances varying from the mid-infrared to the far infrared spectral range [61]. Post-growth patterning of quantum dots and transferring of the pattern to the semiconductor layer by etching was one of the earliest implemented methods [62]. Another method is selective growth of a compound semiconductor with a narrower bandgap on the surface of another semiconductor with a wider bandgap.

These techniques have been successfully applied to study the transport properties of quantum dots [63]. First observations of intersublevel transitions in the far infrared have been reported in the early 1990s, either in InSb-based electrostatically defined quantum dots [64] or in structured two-dimensional electron gas [65]. In the latter cases, the intersublevel transitions were resonant in the far infrared because of the large lateral sizes of the quantum dots.

One of the most popular ways to realize quantum dots is to rely on a spontaneous formation process. This can be achieved in the so-called Stranski–Krastanow growth mode [66] that occurs during the epitaxial growth between lattice-mismatched semiconductors. The quantum dot formation is driven by the strain accumulated in the epitaxial layer. Above a given critical thickness of the wetting layer (WL), islands with scales ranging in the nanometer range are spontaneously formed. This growth mode can be successfully obtained by different growth techniques, including molecular beam epitaxy, chemical vapor deposition, or metal–organic chemical vapor deposition. To date, the most studied type of self-assembled quantum dots is the InAs/GaAs system, which can be considered as a model system [67]. An atomic force microscope image of the InAs island size distribution is shown in Fig. 2.23 [15].

However, since the key parameter which governs the growth is the lattice-mismatch between the semiconductors (7% in the case of InAs on GaAs), self-assembled quantum dots can be easily obtained with other materials like germanium and silicon [68]. The interest for self-assembled quantum dots relies not only on the easy formation process but also on their potential applications

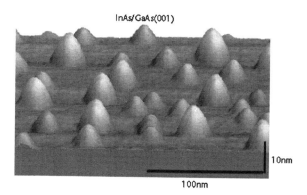

Fig. 2.23 An atomic force microscope image of InAs islands on GaAs(001) [15]

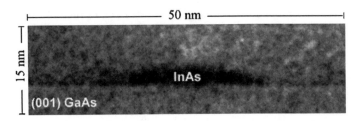

Fig. 2.24 Cross section image of an InAs quantum dot embedded into a GaAs matrix obtained by transmission electron microscopy [62]

due to their compatibility and their possible integration with standard III–V and IV–IV electronics.

The shape, size and compositions of the quantum dots can be significantly modified during regrowth or post-growth annealing or by applying complex growth sequences [15]. In the case of the InAs/GaAs heterosystem, self assembled quantum dots with various shapes have been reported in the literature; including lens shaped quantum dots [69] square-based pyramids [70] or elongated pyramids [71]. Different aspect ratios, defined as the height divided by the base or diameter, have also been reported. The lens-shaped quantum dots are usually rather flat, with an aspect ratio of the order of 1:10, with typical dimension of 2.5 nm height and 25 nm diameter [69]. The pyramids exhibit facets with higher index planes and higher aspect ratios (1:2). Transmission electron microscopy (TEM) structural characterizations show that the InAs dots grown on GaAs roughly have a lens-shaped geometry with a low aspect ratio of ~ 0.1 as seen in Fig. 2.24 [61].

The composition, the strain–relaxation, the interdiffusion and the segregation are strongly dependent on the growth parameters, like the temperature, the ratio between the III–V elements fluxes, the capping procedure etc. It is therefore very difficult to compare quantitatively the results which have been reported in the literature so far. The electronic spectrum of the dots is dependent on all these parameters, and in turn is expected to vary from one system to another.

One drawback of the self-assembled quantum dots as compared to lithography-defined quantum dots is their size distribution, which can put a limitation for device application. The size dispersion, which is inherent to the Stranski–Krastanow growth mode formation, is generally of the order of a few percent around the mean size. The mean size can evidently be varied during the growth process, which opens in turn a route to tune the energy of the confined energy levels [61].

Multiple-layer stacking of islands is often required when implementing QDs in the active region of optical devices to obtain enough interaction between the confined electrons and the electromagnetic field. In this case, a spacer layer determines the size and density of QDs. Below a certain thickness of the spacer (GaAs in this example), the islands align themselves according to the bottom layer and their size increases as we move toward the upper layers. Increasing the thickness of the spacer may result in independent growth of islands with nearly constant size and density in different layers [15].

However another route to tailor the electronic spectrum of the dots is to rely on the electronic coupling that can be achieved by vertical coupling of the quantum dots. This vertical coupling is possible due to the vertical alignment driven by the strain-field induced by one quantum dot. Electronic coupling between stacked layers provides more flexibility for wave function engineering in quantum dots, as it is routinely achieved with quantum wells. The stacking of quantum dots corresponds to the transition between artificial atoms to artificial molecules [61].

Intersublevel transitions offer also great opportunities for the development of new devices in the mid-infrared. By analogy with quantum well intersubband photodetectors, one of the first device applications that have been proposed was the realization of quantum dot infrared photodetectors for infrared detection and imaging. Clearly, one of the anticipated advantages of using quantum dots instead of quantum wells is related to the difference in capture and relaxation mechanisms between the dots and the wells. This should result in an improved responsivity, a lower dark current and a higher operation temperature. Besides, the polarization selection rule for intersublevel absorption differs between quantum dots and quantum wells. Normal incidence operation can therefore be expected with quantum dots. Quantum dots allow also the study of original structures and geometries like lateral quantum dot photodetectors which cannot be investigated with quantum wells [15].

2.5.2 An Overview of Quantum Dots Photodetectors

Detection of terahertz wave like any other electromagnetic radiation may be done by coherent or incoherent techniques. Coherent detection systems such as Schottky diode mixers, nonlinear optical crystals or coatings and gated photoconductive antennas, or switches usually accomplished along with frequency conversion. The coherent techniques despite of good sensitivity need a high degree of sophistication and instrumentation. Incoherent detectors involve heat based devices such as bolometers or those made with pyroelectric crystals. These types of detectors are generally slow and operate at low temperatures. Detectors based on Semiconductor materials (such as doped Ge detectors [72] and photoconductive detectors triggered by femtosecond optical pulses [73]) and heterojunction (such as heterojunction interfacial work function internal photoemission detectors [74], and high-electron mobility transistors operating in the plasma-wave regime [75]) have been also studied and characterized. More recently, quantum-confinement based detectors [76] consist of quantum-well infrared photodetectors (QWIPs) and quantum dot infrared photodetectors (QDIPs) have generated interest. It was observed that a QDIP consisting of a multilayered self-organized In(Ga,Al)As/Ga(Al)As quantum dot active region can detect a broad range of infrared (IR) wavelengths [77–80]. Mid-infrared photoconductivity at around 3 μm has also been studied for delta-doped InAs/AlGaAs quantum dots for subbands to continuum transitions [81, 82].

The advantages of QDIPs result from three-dimensional carrier confinement in quantum dots. The associated advantages include:

(1) Intrinsic sensitivity to normal incidence light.
(2) Long lifetime of photoexcited electrons due to reduced electron–phonon scattering.
(3) Low dark current due to three-dimensional quantum confinement and reduced thermionic emission.

Intersubband absorption in zero-dimensional quantum dot structures has advantages in optical applications compared with two-dimensional quantum well structures. This is due to their sharp delta-like density of states, the reduced intersubband relaxation times and so lower detector noises in these nanostructures [81, 82]. Intersubband absorption of GaAS-based quantum dot structures has been extensively investigated in recent years. For example, infrared absorption has been reported for charged InGaAs quantum dots for wavelengths higher than 20 µm, and for doped InAs dots in the range of 10–20 µm, respectively. Long-wavelength infrared detection is one of the major applications of self-assembly grown semiconductor quantum dots [83]. Most of long-wavelength infrared detectors are generally limited to the peak position wavelength range of 4–9 µm [84, 85]. A study of the intersubband absorption in InAs/GaAs quantum dots has been done in [84]. The long-wavelength infrared intersubband absorption in $In_{0.3}Ga_{0.7}As$/GaAs multiple quantum dots is reported in [83]. With variation of the number of monolayers (between 10 and 60), the peak position wavelength varies in range $\lambda \sim$ 8.6–13 µm. While these QDs have been demonstrated successfully in mid-infrared wavelength photodetectors, the promise of new applications at longer wavelengths in the far-infrared (30–300 µm) or terahertz (1–10 THz) region of the spectrum is providing motivation to extend their operating wavelength [86].

The first observation of mid-infrared photoconductivity with InAs quantum dots was reported in 1997 [61, 81]. Far infrared photoconductivity (17 µm) in self-organized InAs quantum dots was reported in 1998 [87]. The photoresponse remained however very noisy due to the high temperature of the measurement (90 K) and could not be observed for a bias larger than 0.5 V. A short wavelength InGaAs photodetector with InGaP barriers is investigated by Kimet et al. [88]. The photoconductivity was measured at normal incidence. At the peak wavelength of 5.5 µm, the responsivity was 130 mA/W and the detectivity was 4.74×10^7 cm $Hz^{1/2}$/W at 77 K. This figure of merit remains however lower more than two orders of magnitude as compared to GaAs/AlGaAs QWIPs operating at the same wavelength. The reported InAs/GaAs QDIP in [89] has peak at 10 µm and operates at normal incidence. A peak detectivity of 7×10^9 cm $Hz^{1/2}$/W is achieved at 30 K. This value remains much weaker than the achieved value with state of the art quantum well infrared photodetectors. The detectivities of 6×10^8, 5×10^8 cm $Hz^{1/2}$/W are obtained at room temperature and 80 K, respectively at 9 µm for far infrared photodetector using self-assembled InAs quantum dots [90].

Several groups have reported some results for QDIPs with AlGaAs barriers. Liu et al. have embedded 50 layers of InAs quantum dots with $AL_{0.33}Ga_{0.65}As$ barriers. The reported results have shown that the AlGaAs barriers induce blue shift. At 80 K, a responsivity of 0.1A/W has observed at 5 μm and bias voltages of ∼3 V [91]. Another approach consists on using an $Al_{0.3}Ga_{0.7}As$ current blocking barrier between the contacts and the active region where detectivity values, $D^* \sim 3 \times 10^9$ cm $Hz^{1/2}$/W at 100 K, were measured for a peaked photoresponse around 3.75 μm [92, 93]. Using AlGaAs blocking barriers, a detectivity of $D^* \sim 10^{10}$ cm $Hz^{1/2}$/W at 77 K with a photoresponse peaked at 6.2 μm and a 0.7 V bias is reported with responsivity of 14 mA/W [94]. The same authors have reported an InAs QDIP that utilize $In_{0.15}Ga_{0.85}As$ strain–relief cap layers [95]. This device exhibit normal-incidence photoresponse peaks at 8.3 or 8.8 μm for negative or positive bias, respectively. At 77 K and −0.2 V bias, the responsivity is 22 mA/W and the peak detectivity D^* is 3.2×10^9 cm $Hz^{1/2}$/W. The highest responsivities have been achieved using lateral quantum dot infrared photodetectors where the carrier transport is shifted to a neighboring channel with high electron mobility. Lee et al. [96] have reported a responsivity of 4.7 A/W at low temperature (10 K) for a 9 V applied bias for the first time.

Recently, Chu et al. [97] have demonstrated an 11 A/W responsivity associated with a resonant photoresponse around 186 meV (6.65 μm) using an InGaAs channel layer. A Ge quantum dot photodetector has been demonstrated using a MOS tunneling structure [98]. The responsivities of presented photodetector in the case of five-period Ge quantum dot are 130, 0.16 and 0.08 mA/W at wavelengths of 820, 1,300 and 1,550 nm, respectively. The device with 20-period Ge quantum dot introduces the responsivity of 600 mA/W at 850 nm and the reported room temperature dark current density is 0.06 mA/cm^2. The optimized growth of multiple (40–70) layers of self-organized InAs quantum dots separated by GaAs barrier layers in order to enhance the absorption of quantum-dot infrared photodetectors (QDIPs) is investigated in [80]. In devices with 70 quantum-dot layers, at relatively large operating biases (smaller than 1.0 V), the dark current density and the peak responsivity are 10^{-5} A/cm^2 and ∼0.1–0.3 A/W measured for temperature ranges 150–175 K, respectively. The peak detectivity varies in the range of $(6 \times 10^9$–10^{11} cm $Hz^{1/2}$/W) for temperature range (100–200 K).

A resonant tunneling quantum-dot infrared photodetector has investigated theoretically and experimentally in [12]. In this device, the transport of dark current and photocurrent are separated by the incorporation of a double barrier resonant tunneling heterostructure for each quantum-dot layer. The proposed device uses $In_{0.4}Ga_{0.6}As$–GaAs quantum dots and has implemented using molecular beam epitaxy. The introduced system was designed to operate at room temperature and 6 μm. Also the measured data exhibit a strong photoresponse peak at 17 μm. The dark current in the tunneling based devices are almost two orders of magnitude smaller than those in conventional devices. Measured dark current values are 1.6×10^{-8} A/cm^2 at 80 K and 1.55 A/cm^2 at 300 K for 1 V applied bias. Measured values of peak responsivity and specific detectivity are 0.063 A/W and 2.4×10^{10} cm $Hz^{1/2}$/W, respectively, under a bias of 2 V, at 80 K for the 6 μm response. For the 17 μm response, the measured values of peak responsivity and detectivity at 300 K are 0.032 A/W and 8.6×10^6 cm $Hz^{1/2}$/W under 1 V bias.

In In(Ga)As/GaAs quantum dots, the intersublevel energy spacings or the energy difference between the dot and continuum states is normally 40–60 meV, which corresponds to the mid-IR and FIR wavelength ranges. In the other words, the upper cutoff wavelength for detection with QDIPs is limited to less than 25 μm [79]. Therefore, the dot heterostructure and/or the dot size need to be engineered for detection at longer wavelengths and in the terahertz range. Bhattacharya and his coworkers [99] have reported the performance characteristics of tunnel QDIPs, incorporating $In_{0.6}Al_{0.4}As$/GaAs self-organized quantum dots of reduced size in the active region, which exhibit spectral response with peak and cutoff wavelengths of 50 and 75 μm (~ 4.0 THz), respectively. The conduction band diagram of an $In_{0.6}Al_{0.4}As$/GaAs quantum dot layer and the associated resonant tunnel heterostructure for this structure are shown in Fig. 2.25a.

A single 60 Å thick $Al_{0.1}Ga_{0.9}As$ barrier is incorporated before each dot layer to form a quantum well with well-defined final states for the photoexcited electrons.

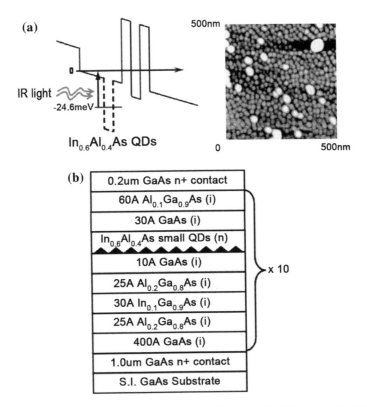

Fig. 2.25 **a** Single period conduction band schematic diagram and AFM image of In0.6Al0.4As/GaAs dots; **b** schematic heterostructure of T-QDIP grown by molecular beam epitaxy [99]

2.5 Quantum Dots THZ-IR Photodetector

The width of the well region and the composition of the barrier can be varied to tune the final states in resonance with the resonant state of the double barrier heterostructure. For detection of terahertz radiation, the energy spacing between the confined state in the dot and the quasibound states in the well has to be of the order of 10 meV or less. This transition is illustrated in Fig. 2.25a. To achieve this, the $In_{0.6}Al_{0.4}As/GaAs$ quantum dots have been grown in the active region of the devices, instead of the more conventional InAs dots. Incorporation of Al into the dot material serves two purposes. First, due to the larger band gap of InAlAs, compared to InAs, the bound state energies are closer to the GaAs barrier energy, and hence to the quasibound states in the well. Second, due to the smaller migration rate of Al atoms on the growing surface during epitaxy, the Al-containing islands (dots) are smaller in size compared to InAs dots and the dot confined states are higher in energy.

In this study, the density of $In_{0.6}Al_{0.4}As$ dots ($\sim 3 \times 10^{11}$ cm^{-2}) is generally an order of magnitude larger than that of InAs dots, which helps to absorb more of the incident radiation. The schematic heterostructure of T-QDIP is shown in Fig. 2.25b.

The energies of the bound states are indicated in Fig. 2.25a. The quantum dots are also doped with Si such that the bound states are occupied.

The dark current density of the device, with the smaller sized InAlAs dots as a function of bias voltage and temperature, is shown in Fig. 2.26. The dark current densities at a bias of 1 V are 4.77×10^{-8}, 2.03×10^{-2}, and 4.09 A/cm^2 at 4.2, 80, and 150 K, respectively. These values are very low compared to other terahertz detectors [74, 100]. This low dark current density is due to the existence of the double barrier tunnel heterostructure. For comparison, the dark current densities in a device with larger sized dots, measured at 80 K, are also included. It is apparent that devices with larger dots are more suitable for high temperature operation.

The spectral response of the tunnel-QDIP with smaller dots at 4.6 K, with bias of 1.0 V, is shown Fig. 2.27a. The peak responsivity is about 0.45 A/W and the wavelength corresponding to this peak is around 50 µm which agrees with the

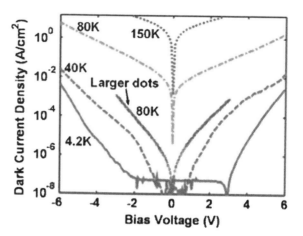

Fig. 2.26 Measured dark current density as a function of bias and temperature [99]

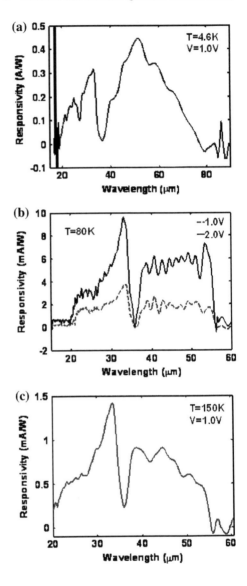

Fig. 2.27 Measured spectral responsivity of T-QDIP at a 4.6 K, b 80 K, and c 150 K under bias of 1 V [99]

calculated energy difference between the QD bound state and the quasibound state in the well of 24.6 meV (50.4 μm). The cutoff wavelength is ∼75 μm, which corresponds to ∼4.0 THz. The transition between the dot state and the state in the well is expected to be sensitive to normal incidence or s-polarized radiation. This has been verified earlier in QDIPs [79]. In the dot-well system, the states in the well are no longer z confined, but also have a radial component. The dark region (dip) in the spectral response centered at (36 μm) is due to longitudinal optical phonon absorption in GaAs. This same phenomenon has been observed by other GaAs based detectors [74, 101]. The spectral response appears to be fairly broad.

The transition is believed to be from the dot bound states to quasibound states in the well and the spectral width of such transitions will not match the observed full width at half maximum (FWHM) of ~35 μm, which corresponds to 23 meV. The observed linewidth is attributed to size nonuniformity of the self-organized dots which give rise to linewidths of ~30–40 meV in the interband photoluminescence spectra. Figure 2.27b, c show responsivity spectra at higher temperatures from a device with the larger sized $In_{0.6}Al_{0.4}As$ dots. The long-wavelength response is shifted to shorter wavelengths. The device can be operated at a temperature of 150 K, which is high compared to other photon-based terahertz detectors. In order to achieve 1–3 THz operation at reasonably high temperature the dot size needs to be reduced, the size uniformity improved, and the tunnel heterostructure needs to be further optimized to keep the dark current low. The dot size can be reduced by increasing the Al content in the dots and by reducing the growth temperature.

The specific detectivity (D^*) of the devices at different temperatures and applied biases is obtained from the peak responsivity R_p and noise density spectra S_i. The latter is measured with a dual channel fast Fourier transform (FFT) signal analyzer, which displays a FFT spectrum of voltage versus frequency, and a low noise preamplifier. A thick copper plate is used as a radiation shield to provide the dark conditions for the measurements. The value of D^* is calculated from:

$$D^* = R_p A^{1/2}/S_i^{1/2} (cm\ Hz^{1/2}/W) \qquad (2.66)$$

where A is the illuminated area of the detector. The measured D^* values are 1.64×10^8 and 4.98×10^7 cm $Hz^{1/2}$/W at 4.6 and 80 K, respectively, under a bias of 1 V.

Hofer and his coworkers have done a work [101] with goal of obtaining emission from interdot transitions in the THz region from quantum dots [102] and from quantum dot cascade emitters [103]. They reported a quantum dot photodetector with two response peaks due to interdot transitions and dot-continuum transitions. Two samples have been used for the investigations. One has 30 dot layers (sample A) and other has 20 layers (sample B). The dot layers of both samples are embedded into a GaAs Matrix of 10 nm width which—for sample B—is Si-doped so that each dot is filled with about one electron. Sample B is provided additionally with thin AlAs layers of 1 nm thickness, 1 nm spaced from the dots to restrict the vertical current (Fig. 2.28). From the relatively small distance between the dots the states in the dots must couple vertically [104].

The measurements show a nonlinear asymmetric behavior with turn-on-voltages of 0.5 and −0.14 V for sample A and 0.66 and −2.5 V for sample B. The dark current is of the order of mA. Sample B shows a smaller dark current than sample A due to the AlAs barriers. The 300 K photoluminescence data-depicted in Fig. 2.29 show for both samples sharp peaks at 994 meV and at 1,021 meV with a FWHM of 57 and 54 meV whereas the intensity for sample B is 7 times higher than for sample A. This can be attributed to the AlAs barriers which improve the carrier capture.

Fig. 2.28 Schematic band diagram showing the structure of the samples [101]

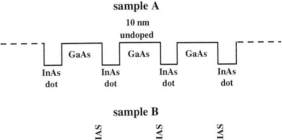

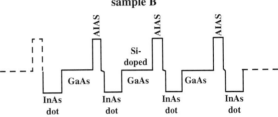

Fig. 2.29 Room temperature photoluminescence spectrum. The peaks show the ground state dot recombination [101]

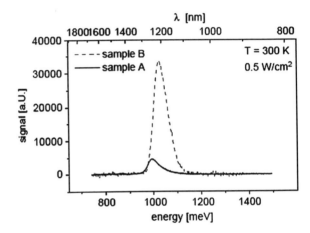

The spectral dependence of the photoresponse of the detector structure is measured using a standard FTIR spectrometer with a glow-bar infrared source. The samples are measured in normal incidence geometry. In Fig. 2.30 the measured transition from the ground state of the quantum dots to the continuum of samples B and A is depicted. As it is expected the higher transition energy for sample B compared to sample A can be seen since the electrons have to be excited into the higher lying lowest miniband of the AlAs–GaAs superlattice.

In fact, there is an energy difference of 27 meV between the peaks of the two samples. This is in exact agreement with the calculation of the miniband dispersion relation done in the envelope function approximation [105]. The lowest energy level of the first miniband lays 27 meV above the GaAs band edge. Sample B shows also a large signal at zero bias which we ascribe to the built in field between the ionized donors next to the dots and the electrons in the dots. The GaAs matrix of sample A is not doped. Therefore there is no photoresponse signal at the

2.5 Quantum Dots THZ-IR Photodetector

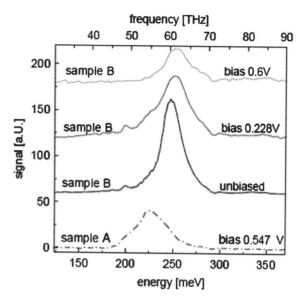

Fig. 2.30 Photoresponse signal of samples A and B due to ground state-continuum transitions in the quantum dots at different bias voltages, T = 10 K [101]

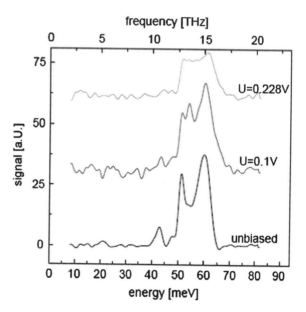

Fig. 2.31 Photoresponse signal of sample B due to interdot transitions in the quantum dots at different bias voltages, $T = 10$ K [101]

unbiased state. Figure 2.30 shows for sample B that the signal is getting smaller at higher bias voltages. This is due to the higher dark current which makes the signal smaller. Measurements in a lower frequency range show a broad peak with several features around 55 meV—Fig. 2.31—which can be interpreted as interdot transitions (see also Pan et al. [106]).

The strong normal incidence response supports the interpretation that the observed peaks are from optical transitions in the quantum dots rather than from the two-dimensional wetting layer. Sample A does not show this peak due to the larger dark current in this sample. In contrast to Chu et al. [107] and Pan et al. [106], in this report interdot transitions and bound to continuum transitions for the same sample possible due to the GaAs–AlAs superlattice miniband which reduces the dark current and noise of the detector [108, 109].

For detecting virus, explosives and bio-images with their molecular vibration frequency locating at the 0.1–30 THz frequency range [110–114], a detector which can detect far infrared wavelength is necessary. Although QDIPs have advantages on detecting infrared light, the detection range is limited to the middle infrared (from several micrometers to tens of micrometer). The quantum ring infrared photodetectors (QRIPs) have the potential of high response speed, longer life time, low dark and noise current, and much better three-dimensional confinement than those of QDIPs; it also exhibits the far infrared wavelength detection ability [115]. Lee and his coworkers [116] have demonstrated the In(Ga)As quantum ring terahertz photodetector which has a cutoff wavelength at 175 μm.

In this work QDs and QRs are grown on semi-insulating (100) GaAs substrate using VG V80H MKII solid source MBE equipped with valved craker sources under As$_2$ beam. The structure consisted of 800 nm n+ GaAs/50 nm GaAs/2ML InAs QD annealing at 520°C for 10 s/1.14 nm GaAs capping layer/50 nm GaAs/400 nm n+ GaAs (Fig. 2.32). After two steps annealing (Fig. 2.33), the QR structure is completed (Fig. 2.34). The first annealing step of InAs QD is to make the QDs with uniform size and density. The second annealing step after the deposition of GaAs capping layer is to let the In atoms out-diffuse from the central QDs and form the QRs.

The thickness of the GaAs capping layer is the important factor for the QRIPs detection wavelength [116].

Figure 2.35 shows PL spectra of various samples with different GaAs capping layer thickness. The thinner the capping layer, the closer the PL peak energy is to the GaAs bandgap. It means that there is opportunity to detect the long wavelength infrared radiation.

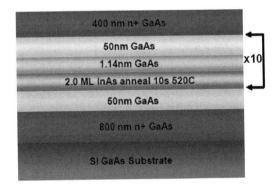

Fig. 2.32 QRIP device structure [116]

2.5 Quantum Dots THZ-IR Photodetector

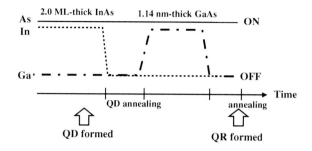

Fig. 2.33 Two steps annealing process to form the quantum ring [116]

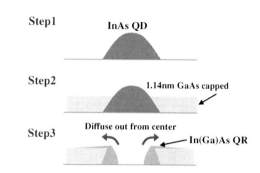

Fig. 2.34 Formation of the quantum ring from the quantum dot [116]

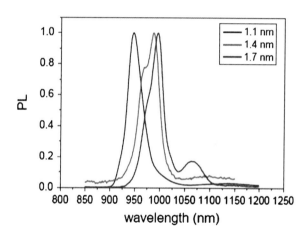

Fig. 2.35 PL Spectra of QD with different GaAs capping layer thickness [116]

Figure 2.36 shows the responsivity of QRIP at a bias of 80 mV and temperature 8 K. It is very clear from the figure that the QRIPs can detect infrared signal at three ranges, i.e., 45–75, 75–100 and 100–175 μm.

The overview presented above about QDIP shows that there are several challenges in terahertz spectra photodetection which can be categorized as follow:

1. In general, the detection range of QDIPs is limited to mid-infrared.
2. In order to achieve terahertz operation in dot-well system, the dot size needs to be reduced.

Fig. 2.36 Repsonsivity of QRIP with 1.14 nm GaAs capping layer thickness [116]

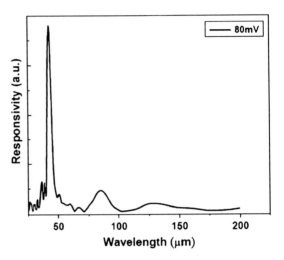

3. Using the 2D states in quantum terahertz photodetection systems leads to decreasing the optical gain and so responsivity.
4. The practical realization of terahertz quantum dot photodetector needs some efforts to be done.

2.5.3 Ultra-High Detectivity Room Temperature THz-IR Photodetector Based on Resonant Tunneling Spherical Centered Defect Quantum Dot

There are several interesting published works to overcome the existing problems in terahertz and long wavelength photodetection. One of the important features of these devices is the ability of detectors in operating at room temperature. Unfortunately, there have not been a suitable proposal for working in room temperature and holding acceptable other characteristics until now. For this reason in this part we describe a quantum dot-based structure for operating at room temperature as well as other high-level characteristics.

According to traditional quantum size effect idea, operating in intersubband-long-IR wavelengths requires large size of quantum structure which leads to low sheet density of quantum dots in each layer in optoelectronic devises. On the other hand, the absorption peak weakens when the resonant frequency is shifted to lower energies (long wavelengths). The maximum reported value for absorption coefficient is about 1.7×10^4 cm^{-1} at 45 μm resonant wavelength [117]. It is obvious that these problems degrade device performances [118]. So it will be interesting to obtain long wavelength transition resonances without increasing the size of the quantum structure and increasing of the absorption peak in lower energies [119]. We show that with introducing a defect in center of quantum dot, it will be

2.5 Quantum Dots THZ-IR Photodetector

possible to increase the absorption coefficient of the proposed structure without increasing the dot size [119–123].

The next motivation of this section is to explore the possibility of extending the detection wavelength of QDIP to terahertz wavelengths. Improving the QDIPs' performance depends fundamentally on minimizing the leakage (dark) current which plagues all light detectors. Three major factors contribute to the existence of the dark current. First factor is the sequential tunneling from between quantum wells through the barrier layer. The tunneling actually has to be mediated by a 'third party', such as a phonon or another electron and is fairly independent of temperature and is thought to dominate below 30 K. Especially in the quantum dot structures, this term is much negligible due to phonon bottleneck effect. Secondly, there is thermally assisted tunneling or field induced emission, which involves thermal excitation within the well followed by tunneling into the continuum. The final contribution is called thermionic emission in which there is direct excitation into the continuum band. It is found that both of the sequential tunneling and the thermionic emission contributions to the dark current increase as the wavelength of the detector extends from the mid- to the far-infrared [86, 124].

In conventional bound-continuum terahertz photodetectors, the electron energy level will be closer to the top of the quantum well. In fact, the energy of the incoming photons may be of the order of the thermal broadening of the electron distribution. Therefore all the mentioned contributions to the dark current may be expected to increase [86, 124].

Intersubband transitions in GaN-based hetrostructures have been the topic of extensive researches for their advantages. Broad wavelength range and high-temperature operation are available in these structures [10]. In this part, we introduce a GaN-based resonant tunneling spherical centered defect quantum dot (RT-SCDQD) to increase the responsivity, decrease the dark current and enhancement of the detectivity in THz range. It will be shown that the responsivity increases due to increasing the absorption coefficient in SCDQD structure. Since the deep intersublevel transitions, which are achievable in wide conduction band offset material (GaN/AlGaN), increase the activation energy, so the second and third terms of the dark current are going to be decreased. In order to collect the electrons from deep excited level, a double barrier structure, which resonances with this level is jointed to the quantum dot structure. This structure cancels the escape of ground state electrons through the tunneling, leading to ultra small ground state dark current.

2.5.3.1 Centered Defect Quantum Dot-Based Photodetector Structure and Simulations

The introduced basic cell structure, resonant tunneling spherical centered defect quantum dot (RT-SCDQD), for room temperature THZ IR-photodetector is illustrated in Fig. 2.37a [125]. In this structure, we consider a spherical quantum dot with radius b. Then a spherical defect with radius a inserted in center of dot

Fig. 2.37 Potential distribution of RT-SCDQD, **a** 3-D scheme and **b** potential distribution [125]

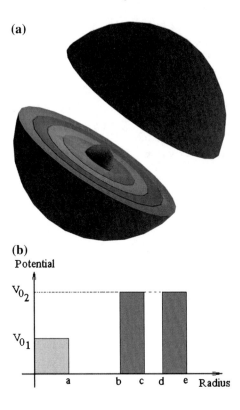

[119]. Finally a resonant tunneling double barrier is attached to the system. Also, for mathematical evaluation of the structure potential distribution of the introduced structure is illustrated in Fig. 2.37b. The introduced structure is based on AlGaN/GaN heterostructure.

The introduced centered defect quantum dot structure (Fig. 2.38a) may be implemented with a method based on current implementation technologies (MBE, Ion implantation and Stranski–Krastanov) [126]. In this method, first, a thin layer with thickness of 2× dot radius (GaN) is grown on the substrate by MBE (Fig. 2.38b). Then a ∼200 nm × 200 nm window is introduced on grown dot layer by masking technique. Defect material particles (Ga + N + Al) are bombarded to considered defect region in dot by ion implantation technique. The temperature in annealing form is applied for defect construction. By applying controlled temperature and due to the inherent strain between layers and substrate, the layers can become separated as quantum dot islands just like Stranski–Krastanov mode (Fig. 2.38c).

In this section mathematical formulation for description of the electrical and optical properties of RT-SCDQD is presented. For this purpose effective mass approximation in spherical coordinate is considered. The Schrödinger equation in the case of slowly varying envelope approximation in spherical coordinate is given as follows.

2.5 Quantum Dots THZ-IR Photodetector

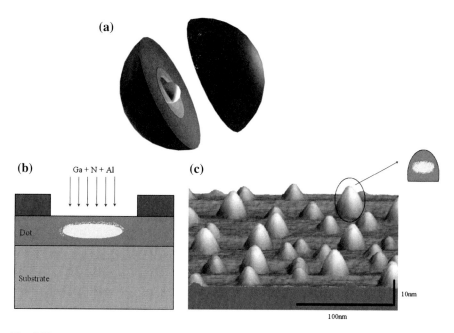

Fig. 2.38 **a** SCDQD structure, **b** Ion implantation of defect, and **c** forming of defected quantum dot islands [119, 126]

$$\left\{-\frac{\hbar^2}{2m_i^*}\left[\frac{1}{r}\frac{\partial^2}{\partial r^2}r + \frac{1}{r^2\sin\theta}\frac{\partial}{\partial\theta}\left(\sin\theta\frac{\partial}{\partial\theta}\right) + \frac{1}{r^2\sin^2\theta}\frac{\partial^2}{\partial\phi^2}\right] + V_i(r)\right\}\Psi(r,\theta,\phi)$$
$$= E\Psi(r,\theta,\phi) \tag{2.67}$$

where

$$m_i^* = \begin{cases} m_0^* & 0<r<a \\ m_1^* & a<r<b \\ m_2^* & b<r<c \\ m_3^* & c<r<d \\ m_4^* & d<r<e \\ m_5^* & e<r \end{cases} \quad \text{and} \quad V_i(r) = \begin{cases} V_{01} & 0<r<a \\ 0 & a<r<b \\ V_{02} & b<r<c \\ 0 & c<r<d \\ V_{02} & d<r<e \\ 0 & e<r \end{cases}$$

are effective mass and potential distribution, respectively. Using the standard method based on separation of variables the following solution for wave function is proposed.

$$\Psi_{n\ell m} = R_{n\ell}(r)Y_{\ell m}(\theta,\phi) \tag{2.68}$$

Considering the proposed solution for wave function, the angular dependent term as spherical harmonics is given in the following [127].

$$Y_{\ell m}(\theta, \phi) = (-1)^m \left[\frac{(2\ell+1)(\ell-m)!}{4\pi(\ell+m)!} \right]^{\frac{1}{2}} P_\ell^m(\cos\theta) e^{im\phi}, \qquad (2.69)$$

After substituting of Eq. 2.68 into 2.67 and mathematical manipulation the radial part of the wave function is solution of the following differential equation.

$$r^2 \frac{d^2 R}{dr^2} + 2r \frac{dR}{dr} + \left\{ \frac{2m_i^*}{\hbar^2} [E - V_i(r)] r^2 - \ell(\ell+1) \right\} R = 0. \qquad (2.70)$$

Solution of Eq. 2.70 for $E < V_{01}$ is given as:

$$R = \begin{cases} [C_{01} i_l(k_0 r)] \sqrt{\frac{2}{\pi}} & 0 < r < a \\ [C_{11} j_l(k_1 r) + C_{12} n_l(k_1 r)] \sqrt{\frac{2}{\pi}} & a < r < b \\ [C_{21} i_l(k_2 r)] \sqrt{\frac{2}{\pi}} + [C_{22} k_l(k_2 r)] \sqrt{\frac{2}{\pi}} & b < r < c \\ [C_{31} j_l(k_3 r) + C_{32} n_l(k_3 r)] \sqrt{\frac{2}{\pi}} & c < r < d \\ \left[C_{41} i_l(k_4 r) \sqrt{\frac{2}{\pi}} + C_{42} k_l(k_4 r) \right] \sqrt{\frac{2}{\pi}} & d < r < e \\ [C_{51} j_l(k_5 r) + C_{52} n_l(k_5 r)] \sqrt{\frac{2}{\pi}} & e < r \end{cases}, \qquad (2.71)$$

Also, for $E < V_{01}$, we have:

$$R = \begin{cases} [C_{01} j_l(k_0 r)] \sqrt{\frac{2}{\pi}} & 0 < r < a \\ [C_{11} j_l(k_1 r) + C_{12} n_l(k_1 r)] \sqrt{\frac{2}{\pi}} & a < r < b \\ [C_{21} i_l(k_2 r)] \sqrt{\frac{2}{\pi}} + [C_{22} k_l(k_2 r)] \sqrt{\frac{2}{\pi}} & b < r < c \\ [C_{31} j_l(k_3 r) + C_{32} n_l(k_3 r)] \sqrt{\frac{2}{\pi}} & c < r < d \\ \left[C_{41} i_l(k_4 r) \sqrt{\frac{2}{\pi}} + C_{42} k_l(k_4 r) \right] \sqrt{\frac{2}{\pi}} & d < r < e \\ [C_{51} j_l(k_5 r) + C_{52} n_l(k_5 r)] \sqrt{\frac{2}{\pi}} & e < r \end{cases}, \qquad (2.72)$$

where

$$\begin{cases} \kappa_0 = \sqrt{\frac{2m_0^*(V_{01}-E)}{\hbar^2}} & 0 < r < a \\ \kappa_0 = \sqrt{\frac{2m_0^*(E-V_{01})}{\hbar^2}} & 0 < r < a \\ \kappa_1 = \sqrt{\frac{2m_1^* E}{\hbar^2}} & a < r < b \\ \kappa_2 = \sqrt{\frac{2m_2^*(V_{02}-E)}{\hbar^2}} & b < r < c \\ \kappa_3 = \sqrt{\frac{2m_3^* E}{\hbar^2}} & c < r < d \\ \kappa_4 = \sqrt{\frac{2m_4^*(V_{02}-E)}{\hbar^2}} & d < r < a \\ \kappa_4 = \sqrt{\frac{2m_5^* E}{\hbar^2}} & e < r \end{cases}$$

2.5 Quantum Dots THZ-IR Photodetector

In above equations n, ℓ, and m are integer values. Now, for extraction of the eigenvalues, eigenfunctions and the transmission coefficient of the proposed system, the obtained solution for the Schrödinger equation should satisfy the following normalization and boundary conditions.

$$\int_0^\infty dr\, r^2 R_{n\ell}^2(r) = 1$$

$$\begin{cases} R_{0<r<a}(a) = R_{a<r<b}(a) \\ \frac{1}{m_0^*}\frac{dR_{0<r<a}}{dr}\bigg|_{r=a} = \frac{1}{m_1^*}\frac{dR_{a<r<b}}{dr}\bigg|_{r=a} \end{cases}$$

$$\begin{cases} R_{a<r<b}(b) = R_{b<r<c}(b) \\ \frac{1}{m_1^*}\frac{dR_{a<r<b}}{dr}\bigg|_{r=b} = \frac{1}{m_2^*}\frac{dR_{b<r<c}}{dr}\bigg|_{r=b} \end{cases}$$

$$\begin{cases} R_{b<r<c}(c) = R_{c<r<d}(c) \\ \frac{1}{m_2^*}\frac{dR_{b<r<c}}{dr}\bigg|_{r=c} = \frac{1}{m_3^*}\frac{dR_{c<r<d}}{dr}\bigg|_{r=c} \end{cases}$$

$$\begin{cases} R_{c<r<d}(d) = R_{d<r<e}(d) \\ \frac{1}{m_3^*}\frac{dR_{c<r<d}}{dr}\bigg|_{r=d} = \frac{1}{m_4^*}\frac{dR_{d<r<e}}{dr}\bigg|_{r=d} \end{cases}$$

$$\begin{cases} R_{d<r<e}(e) = R_{e<r}(e) \\ \frac{1}{m_4^*}\frac{dR_{d<r<e}}{dr}\bigg|_{r=e} = \frac{1}{m_5^*}\frac{dR_{e<r}}{dr}\bigg|_{r=e} \end{cases},$$

Now, based on the boundary conditions, we develop and use the transfer matrix method (TMM) to obtain the transmission coefficient for the proposed system. For this purpose, we rewrite the boundary conditions in the form of dynamic matrices as follows

$$\begin{bmatrix} C_{11} \\ C_{12} \end{bmatrix} = P_{1,2} \times \begin{bmatrix} C_{21} \\ C_{22} \end{bmatrix}$$

$$\begin{bmatrix} C_{21} \\ C_{22} \end{bmatrix} = P_{2,3} \times \begin{bmatrix} C_{31} \\ C_{32} \end{bmatrix}$$

$$\begin{bmatrix} C_{31} \\ C_{32} \end{bmatrix} = P_{3,4} \times \begin{bmatrix} C_{41} \\ C_{42} \end{bmatrix} \quad (2.74)$$

$$\begin{bmatrix} C_{41} \\ C_{42} \end{bmatrix} = P_{4,5} \times \begin{bmatrix} C_{51} \\ C_{52} \end{bmatrix},$$

where

$$p_{u,u+1} = D$$

$$\times \begin{bmatrix} \left(n_l'(\kappa_u r) i_l(\kappa_{u+1} r) - \frac{m_u^* \kappa_{u+1}}{m_{u+1}^* \kappa_u} n_l(\kappa_u r) i_l'(\kappa_{u+1} r)\right)\left(\frac{\pi}{2}\right) & \left(n_l'(\kappa_u r) k_l(\kappa_{u+1} r) - \frac{m_u^* \kappa_{u+1}}{m_{u+1}^* \kappa_u} n_l(\kappa_u r) k_l'(\kappa_{u+1} r)\right) \\ \left(\frac{m_u^* \kappa_{u+1}}{m_{u+1}^* \kappa_u} j_l(\kappa_u r) i_l'(\kappa_{u+1} r) - j_l'(\kappa_u r) i_l(\kappa_{u+1} r)\right)\left(\frac{\pi}{2}\right) & \left(\frac{m_u^* \kappa_{u+1}}{m_{u+1}^* \kappa_u} j_l(\kappa_u r) k_l'(\kappa_{u+1} r) - j_l'(\kappa_u r) k_l(\kappa_{u+1} r)\right) \end{bmatrix} \begin{matrix} \text{if } u=1 \Rightarrow r=b \\ \text{if } u=3 \Rightarrow r=d \end{matrix}$$

$$D = 1/\left(j_l(\kappa_u r) n_l'(\kappa_u r) - j_l'(\kappa_u r) n_l(\kappa_u r)\right)$$

for odd values of u and

$$p_{u,u+1} = D$$

$$\times \begin{bmatrix} \left(\frac{\pi}{2}\right)\left(k'_l(\kappa_u r)j_l(\kappa_{u+1}r) - \frac{m^*_u \kappa_{u+1}}{m^*_{u+1}\kappa_u}k_l(\kappa_u r)j'_l(\kappa_{u+1}r)\right) & \left(\frac{\pi}{2}\right)\left(k'_l(\kappa_u r)n_l(\kappa_{u+1}r) - \frac{m^*_u \kappa_{u+1}}{m^*_{u+1}\kappa_u}k_l(\kappa_u r)n'_l(\kappa_{u+1}r)\right) \\ \left(\frac{m^*_u \kappa_{u+1}}{m^*_{u+1}\kappa_u}i_l(\kappa_u r)j'_l(\kappa_{u+1}r) - i'_l(\kappa_u r)j_l(\kappa_{u+1}r)\right) & \left(\frac{m^*_u \kappa_{u+1}}{m^*_{u+1}\kappa_u}i_l(\kappa_u r)n'_l(\kappa_{u+1}r) - i'_l(\kappa_u r)n_l(\kappa_{u+1}r)\right) \end{bmatrix} \begin{matrix} \text{if } u=2 \Rightarrow r=c \\ \text{if } u=4 \Rightarrow r=e \end{matrix}$$

$$D = 2/\left(\pi\left(i_l(\kappa_u r)k'_l(\kappa_u r) - i'_l(\kappa_u r)k_l(\kappa_u r)\right)\right)$$

for even values of u, respectively.

Thus, after multiplication of all matrices obtained in previous relations, one can find the following transfer matrix describing input–output relation of the proposed structure.

$$\begin{bmatrix} C_{11} \\ C_{12} \end{bmatrix} = M \times \begin{bmatrix} C_{51} \\ C_{52} \end{bmatrix}, \tag{2.75}$$

where $M = P_{1,2} \times P_{2,3} \times P_{3,4} \times P_{4,5}$.

In order to calculate the transmission coefficient of resonant double barrier, for description of the forward and backward traveling waves for obtaining of the propagation matrices, we must explain the Bessel Functions interms of the Hanckel forms as follows:

$$\begin{aligned} C_{11}j_l(\kappa_1 r) + C_{12}n_l(\kappa_1 r) &= \alpha_{11}H_l^{(1)} + \alpha_{12}H_l^{(2)} \\ C_{51}j_l(\kappa_5 r) + C_{52}n_l(\kappa_5 r) &= \alpha_{51}H_l^{(1)} + \alpha_{52}H_l^{(2)}, \end{aligned} \tag{2.76}$$

Considering $j_l(x) = \frac{1}{2}\left(H_l^{(1)}(x) + H_l^{(2)}(x)\right)$, $n_l(x) = \frac{1}{2i}\left(H_l^{(1)}(x) - H_l^{(2)}(x)\right)$ [127] and Eq. 2.76 the transfer matrix relation (Eq. 2.75) can be converted as follows:

$$\begin{bmatrix} \alpha_{11} \\ \alpha_{12} \end{bmatrix} = \Gamma \times \begin{bmatrix} \alpha_{51} \\ \alpha_{52} \end{bmatrix}, \tag{2.77}$$

where

$$\Gamma = \left(\frac{1}{2}\right)\begin{bmatrix} (M_{11} + M_{22}) + i(M_{12} - M_{21}) & (M_{11} - M_{22}) - i(M_{12} + M_{21}) \\ (M_{11} + M_{21}) + i(M_{12} - M_{22}) & (M_{11} + M_{22}) - i(M_{12} - M_{21}) \end{bmatrix}.$$

So, the transmission coefficient can be obtained as follows:

$$T(E) = \frac{1}{|\Gamma_{11}|^2}. \tag{2.78}$$

After extraction of the sublevel energies, their corresponding wave functions and the transmission coefficient are obtained. Also, the linear absorption coefficient ($\alpha(\omega)$) for the intersublevel transitions can be clearly calculated by computing the optical susceptibility through the density matrix approach [85, 117, 128] as follows:

2.5 Quantum Dots THZ-IR Photodetector

$$\alpha(\omega) = \frac{4\pi\omega e^2}{V_o \hbar c \varepsilon_0 \sqrt{\varepsilon_r}} \sum_{i,j} |d_{ij}|^2 \times \{f(E_i) - f(E_j)\} \times \frac{\gamma_{ij}}{\gamma_{ij}^2 + (\omega - \omega_{ij})^2} \quad (2.79)$$

where $\omega, e, V_O, c, \varepsilon_0, \varepsilon_r, |d_{ij}| (= |\langle\psi_j|r|\psi_i\rangle|), f(E), \gamma_{ij}(1/\tau_{ij}), \omega_{ij}$ are photon frequency, electron charge, volume of quantum dot, speed of light, permittivity of vacuum, relative permittivity of semiconductor, dipole transition matrix element (electron transition $i \to j$), Fermi–Dirac distribution, relaxation rate (inverse of relaxation time), and transition frequency (resonance frequency between two electronic states), respectively. In above equation Lorentzian broadening is considered [83, 128]. The Fermi energy level is obtained by inverse numerical solution of the following equation [129].

$$N_d = \frac{2}{V_o} \sum_i \{1 + \exp[(E_i - F_c)/k_B T]\}^{-1}, \quad (2.80)$$

where N_d, V_o, E_i, F_c, k_B and T are electron density, volume of spherical dot, sublevel energy, Fermi energy level, the Boltzman constant and temperature, respectively. To determine precisely the Fermi energy level in calculating the above equation, all of the energy levels within a dot should be included.

The simple schematic view of the QDIP based on RT-SCDQD cells is shown in Fig. 2.39. The RT-SCDQD layers are separated by un-doped GaN spacer layers.

The material and structural parameters of the proposed RT-SCDQD based THZ IR-photodetector are given in Table 2.2 [15, 130–132].

In the following, optical and electrical simulated results of the introduced RT-SCDQD based THZ IR-photodetector are presented and discussed. The simulation consists of two parts. First, we investigate the linear absorption coefficient of the introduced SCDQD structure and the effect of parameters of the introduced defect on enhancement of the absorption coefficient is studied. In this study the effect of defect size on the energy levels, wave functions, dipole matrix element and absorption coefficient is discussed. In the second section we describe the performance of the RT-SCDQD photodetector. The effect of defect and double

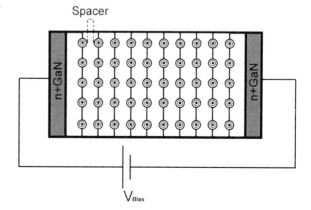

Fig. 2.39 Schematic view of the QDIP structure based on RT-SCDQD [102]

Table 2.2 Material and structural parameters of RT-SCDQD based THZ IR-photodetector [15, 130–132]

$Al_xGa_{1-x}N$ parameters	Unit	Value
Electron effective mass (m^*)	m_0	$0.252x + 0.228$
Band gap ($E_g(x)$)	eV	$6.13x + (1-x) \times 3.42 - x(1-x)$
Band offset ($\Delta E_C(x)$)	eV	$0.7 \times [E_g(x) - E_g(0)]$
Typical Relaxation time (τ)	fs	100
Number density of carriers (N_w)	m^{-3}	1×10^{24}
Electron mobility (μ)	cm^{-2}/Vs	2×10^3
Electron saturation velocity (v_s)	m/s	2×10^5
Recapture life time (τ_r)	s	1×10^{-12}
Surface density of QD(Σ_{QD})	$1/m^2$	1×10^{14}
Band width (Δf)	Hz	1
Radius of detector (r)	μm	200
Relative dielectric constant (ε_r)	–	$8.5x + 10.4(1-x)$
Number of QD layers (N_{QD})	–	10
Capture probability (p_c)	–	0.001
Fill factor (F)	–	0.35

barrier resonant tunneling structure on improvement of responsivity and detectivity is provided.

In Fig. 2.40, we illustrate the effect of defect size on ground and first excited states. The defect mole fraction is considered as a parameter. It is observed that with increasing of the defect size the energy levels are increased but difference between the energy levels is decreased. The illustrated decreasing in difference of energy levels with increasing defect size is related to the fact that the ground energy level is affected more than the first excited level [119]. Finally with increasing of the mole fraction difference between energy levels also is decreased.

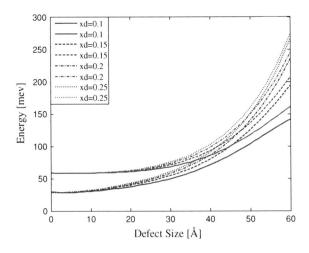

Fig. 2.40 Energy levels (ground and first excited states) versus defect sizes (Å) with different mole fraction of introduced defect ($b = 70$ Å, $xb = 0.3$) [125]

2.5 Quantum Dots THZ-IR Photodetector

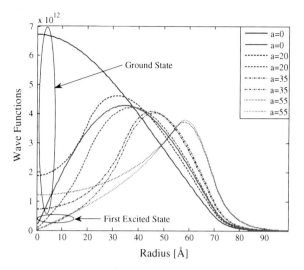

Fig. 2.41 Wave functions (ground and first excited states) versus dot radius (Å) with different defect sizes ($b = 70$ Å, $xb = 0.3$, $xd = 0.15$) [125]

The electron wave function of the structure is illustrated in Fig. 2.41. The wave function inside defect with increasing of defect size is decreased. With increase of defect size, peak of the wave function is shifted to the right hand side within distance between defect and dot. This phenomenon is due to potential barrier effect on electron wave function that is concluded to shifting of the maximum probability of electron [119].

In the Fig. 2.42, the effect of defect size and mole fraction on dipole transition matrix element is investigated. As it is observed with increasing the defect size and also defect height the dipole transition matrix element is increased [119]. Decreasing of the matrix element in large defect size is owing to tunneling leakage

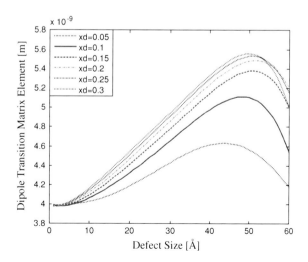

Fig. 2.42 Dipole transition matrix element ($\langle\psi_f|r|\psi_g\rangle$) versus defect width ($b = 70$ Å, $xb = 0.3$) [125]

Fig. 2.43 Absorption coefficient (Ground state → First excited state) versus pump photon energy ($b = 80$, $x_b = 0.3$, $x_d = 0.1$) [125]

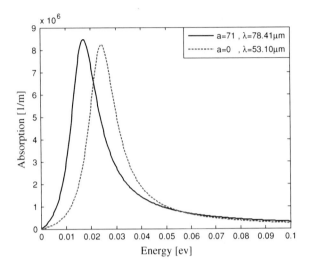

of the wave functions into dot barrier and defect. So, there is an optimum value of defect size for maximum matrix element.

In Fig. 2.43 the effect of defect size on SCDQD absorption coefficient is presented. The absorption coefficient is calculated for ground to first excited state transition. It is shown that with increasing of the defect size, the absorption peak is increased and a red shift is observed. The observed red shift is due to pushing the energy levels up and decreasing the difference between energy levels. As it mentioned before, increasing the defect size increases the dipole transition matrix element and hence increases the peak of absorption coefficient. It is noticeable that the absorption peak increases despite of shifting to lower energies.

Table 2.3 shows the absorption coefficient peak in different defect and dot sizes for the same resonance wavelength. It is found that with the larger defect size, higher absorption coefficient is achievable in small dot size.

As it is observed the SCDQD has large absorption coefficient at long wavelength in the small quantum dot size, thus we use this structure as a basic THZ IR-photodetector cell (Fig. 2.39).

Now based on the proposed basic cell in the following, we present simulated results of the THZ-IR photodetector system. Table 2.4 illustrates the optimal parameters used for obtaining characteristics of the THZ-IR photodetector.

Table 2.3 Absorption coefficient peaks (Ground state → First excited state) for different SCQD-structural parameters ($x_b = 0.3, x_d = 0.1$) [125]

Structure parameters			Absorption(1/m)
$a(\text{Å})$	$b(\text{Å})$	$\lambda(\mu m)$	
55	70	83.53	7.34×10^6
40	70	82.98	6.79×10^6
30	79	82.09	5.68×10^6
20	90	80.62	3.04×10^6
0	100	80.59	1.05×10^6

2.5 Quantum Dots THZ-IR Photodetector

Table 2.4 Design parameters for simulation of THZ IR-photodetector based on RT-SCDQD [125]

Structure parameter	Unit	Value
a	M	55×10^{-10}
b	M	70×10^{-10}
c	M	90×10^{-10}
d	M	110×10^{-10}
e	M	130×10^{-10}
Defect mole fraction (x_d)	–	0.1
Barrier mole fraction (x_b)	–	0.3
Operation temperature	K	83
Spacer layer width	M	130×10^{-10}
Applied voltage	V	2

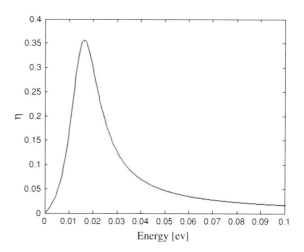

Fig. 2.44 Absorption quantum efficiency of RT-SCDQD based THZ-IR photodetector versus photon energy [125]

Absorption quantum efficiency versus photon energy is illustrated in Fig. 2.44. It is shown that there is a maximum value at 83 μm. It is clear that there is direct relationship between maximum value of the absorption coefficient and the quantum efficiency.

The calculated tunneling probability for the RT-SCDQD basic cell is illustrated in Fig. 2.45. The proposed double barrier parameters are designed such that the resonance energy to be close to first excited state of the quantum dot.

Responsivity of the introduced structure is illustrated in Fig. 2.46. It is observed that the reported large value related to the following factors.

1. For the proposed structure the calculated quantum efficiency is increased considerably due to larger absorption coefficient.
2. Probability of electron escaping is increased in the proposed system owing to decrease of the collection time of electron using resonant double barrier.
3. Optical gain for the proposed photodetector is increased considerably due to inherent properties of quantum dots.

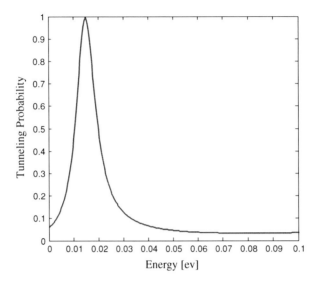

Fig. 2.45 The tunneling probability of RT-SCDQD versus photon energy [125]

Fig. 2.46 Responsivity of RT-SCDQD based THZ-IR photodetector versus photon energy [125]

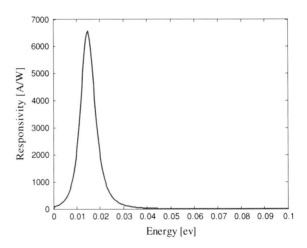

Dark current curves of RT-SCDQD based THZ-IR photodetector versus bias voltage at various temperatures is illustrated in Fig. 2.47. It is shown that with increasing of the temperature the appeared dark current is increased strongly. This increase in the dark current basically related to thermionic effect.

For the proposed system the detectivity is illustrated in Fig. 2.48. The illustrated detectivity is considerable higher than reported values in similar situations.

The calculated detectivity and NEΔT parameters of proposed structure (RT-SCDQD-THZ-IR photodetector) are compared with conventional structure (QD-THZ-IR photodetector) in Table 2.5. It is shown that the proposed complete

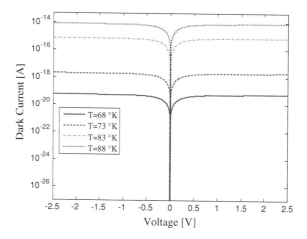

Fig. 2.47 Dark current curves of RT-SCDQD based THZ-IR photodetector versus bias voltage at various temperatures [125]

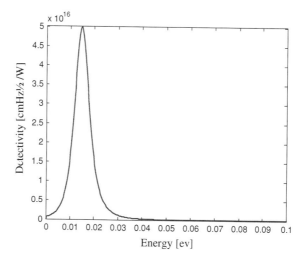

Fig. 2.48 Detectivity of RT-SCDQD based THZ-IR photodetector versus photon energy [125]

structure considerably has large detectivity, low NEΔT and narrow spectra compared the other case. The small line width of escaping probability which is consequence of the resonant tunneling double barrier as it is obvious in Fig. 2.45 yield to narrow line width in RT-SCDQD IR-photodetector detectivity spectra.

The reported ultra high value of detectivity and low value of NEΔT are related to two basic effects. One is responsivity which in the proposed structure is increased due to enhancement of the absorption coefficient in SCDQD compared other quantum dots without defect. Second effect related to decrease of the dark current in the proposed structure. Decreasing of dark current in the proposed structure is done owing to the following reasons.

1. Increasing of the barrier height, concluding to decrease of the thermionic term in the dark current. This subject may be introduce some difficulty in electron

Table 2.5 Calculated detectivity, noise-equivalent temperature difference and full-width at half of the maximum of proposed RT-SCQD_THZ IR-photodetector and conventional QD_THZ IR-photodetector at 83 and 300 K [125]

Structure parameter	D^* [cm Hz$^{1/2}$/W] 83 K 83 μm	D^* [cm Hz$^{1/2}$/W] 300 K 83 μm	NE · T	FWHM (μm)
Conventional QD structure (QD-THZ-RPD)	2.03×10^{10}	5.92×10^7	5.55×10^{-2}	46
Structure (RT-SCDQD-HZ-IRPD)	5×10^{16}	2.29×10^9	1.02×10^{-7}	106

Table 2.6 Design parameters for simulation of proposed RT-SCQD_THZ IR-photodetector and conventional QD_THZ IR-photodetector [125]

Structure parameter	Unit	Proposed QD structure (RT-SCDQD-THZ-IRPD)	Conventional QD structure (QD-THZ-IRPD)
A	M	55×10^{-10}	–
B	M	70×10^{-10}	95×10^{-10}
C	M	90×10^{-10}	–
D	M	110×10^{-10}	–
E	M	130×10^{-10}	–
Defect mole fraction (x_d)	–	0.1	–
Barrier mole fraction (x_b)	–	0.3	0.1
Operation temperature	K	83	83
Spacer layer width	M	130×10^{-10}	40×10^{-10}
Applied voltage	V	2	10

collection in photodetector, which is removed using double barrier element that resonances with first excited state of dot.

2. Using double barrier element in the proposed system, introduces ultra low ground state dark current also.

Finally the proposed photodetector is examined at room temperature. It is observed that the calculated result shows interesting value which illustrates capability for working at room temperature (Table 2.5). This is so interesting and well large value compared previous reported results. The proposed structure illustrates ultra high value compared traditional bound to continuum transition based THZ photodetectors [124]. It can be understood that in this structure we decreased considerably the dark current owing to tuning of intersubband transition to mid conduction band offset to decrease thermal effect (variation of Fermi level and thermionic emission from level to continuum band) and finally extraction and collection of electrons through resonant tunneling double barrier structure. Also, it should be mentioned that the proposed defect in quantum dot increased considerably responsivity of the structure.

Parameters of the proposed structure and conventional structure are given in Table 2.6 for comparison.

2.5 Quantum Dots THZ-IR Photodetector

In this section the proposed structure for basic block of IR photodetectors were evaluated completely and different features investigated. It was shown that the proposed unit cell have excellent advantages for room temperature photodetectors.

2.5.4 Terahertz Photodetector Based on Intersublevel Optical Absorption in Coupled Quantum Dots

In a quantum dot infrared photodetector (QDIP) [77], electronic transitions between energy states can lead to terahertz radiation detection [99]. When the layers are closely spaced in the coupled quantum dot devices, the level splitting takes place due to vertical electronic correlation. This feature plays an important role in determining the electronic and optical properties of multiple self-assembled quantum dots. The multi-confined levels can be explained by the appearance of level splitting due to vertical electronic correlation. There are four important factors that contribute to level splitting: quantum–mechanical coupling (happen when the QDs are stacked closely together), strain (system does not contain a symmetry plane parallel to the base of the QDs and strain does not affect the two QDs in the same way), piezoelectric potentials (origins from the nonzero off-diagonal shear-strain tensor elements) and finally indium migration during the epitaxial growth.

In the field of infrared photodetector, the absorptions in strongly vertically coupled QDs are based on the inter-subband transitions due to the energy level splitting in the same QDs, have many advantages as following. Firstly, two-color inter-subband absorptions based on the energy level splitting can be easily controlled by adjusting the thickness of spacer. Two-color inter-subband absorptions based on the different size of QDs are limited by the random distribution of QDs and it is difficult to control the absorption peaks. Large-size QDs consequentially lead to little full factor and low quantum efficiency. Secondly, the structure of vertically coupled QDs has much stronger normal incidence absorption [133] than that of uncoupled QDs. Thirdly, most devices suffer from low gain or responsivity resulting from the presence of only one quantum dot plane or different size of QDs but these problems could be ameliorated with improving shape and size uniformity of quantum dots and also by adding more quantum-dot planes, so we can use quantum dots planes with coupling in vertical direction. Finally, the coupled quantum dots are too much attractive as the base of the detector structures for detecting the terahertz electromagnetic spectrum [134]. The associated infrared absorption would correspond in this case to an intersublevel transition between the splitting energy levels of vertically coupled dots.

Figure 2.49 shows the schematic view of two typical QDs-stacks that each of them consists of two InAs quantum dots (pyramid or dome shaped) with 0.3-nm-thick wetting layer.

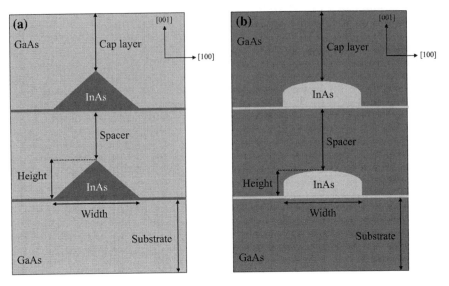

Fig. 2.49 Cross section of the two coupled quantum dot structures with pyramid and dome shaped QDs

In order to theoretically understand the nature of the electronic and optical properties of coupled quantum dots, a first step is to determine the strain distribution in the dot structures and in the matrix material, as the next step, the calculated strain serves as an input to electronic structure calculations. The driving force for vertically self-organized growth is known to be the interacting strain fields induced by the islands. The strain modifies the effective confinement volume in the device, distorts the atom bonds in length and angles, and hence modulates the local band structure and the confined states. The hydrostatic component of strain, for example, usually shifts the conduction and valence band-edges of semiconductors; biaxial strain, on the other hand, modifies the valence bands by splitting the degeneracy of the light- and heavy-hole bands; therefore the strain distribution profoundly affects the electronic and optical properties of the assembly of dots by modifying the energies and wave functions of the electron and hole confined states.

Simulation results show that the top dots increase the strain of the lower dots. This strain increasing is more in dome shaped quantum dots. Because of the special geometry type of dome shaped QD, the upper dot completely cover the lower dot that leads to increasing the coupling effect between quantum dots. Calculated results indicate that the magnitudes of the strain components depend on the geometries and the spacer thicknesses between coupled quantum dots. Because of the presence of strain, the lattice constant is changed and the symmetry of the crystal is reduced, so the bandstructure of semiconductors is generally altered. By taking into account the influence of the strain distributions

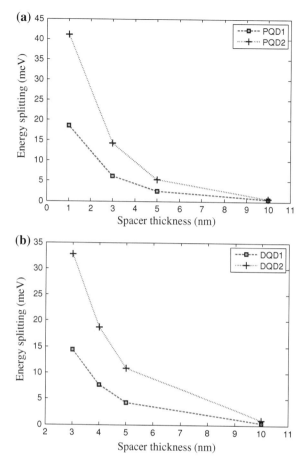

Fig. 2.50 Energy level splitting for **a** pyramid shaped QDs (PQD1: base; 15 nm × hight; 7 nm, PQD2: base; 12 nm × hight; 5 nm). Energy level splitting for **b** dome shaped QDs (DQD1: base; 16 nm × hight; 5 nm, DQD2: base; 13 nm × hight; 3 nm) based structures as a function of spacer thickness

and solving the three-dimensional, effective mass single band Schrödinger equation, the energy levels and wavefunctions in coupled quantum dots may be calculated. Here, only the effects of the quantum–mechanical coupling and strain in energy level splitting have been considered and main focus is about electron ground state splitting. Figure 2.50 shows the level splitting for two different QDs-structures with pyramid or dome shaped QDs as a function of spacer thickness.

It is clear that as the dot separation is narrowed, the dots interact strongly with each other mechanically through the strain field as well as quantum mechanically through wavefunctions overlaps, so the energy level splitting increases. For smaller quantum dots, the coupling effects is more, therefore the levels splitting increase for coupled quantum dots with smaller sizes. The geometry type of quantum dots has straight influence in the rate of energy level splitting. Thanks to high amount of strain field in dome shaped quantum dots, separation increasing between splitted energy levels expected in compared with pyramidal shape dot.

The typical calculated ground state splitting is about $0 \to 41$ meV ($0 \to 10$ THz) and therefore the strongly vertically coupled QDs are promising for fulfill the terahertz range detection due to their energy level splitting. In coupled quantum dots based photodetectors, the sizes of quantum dots, distance between coupled QDs, the geometry types of dots and finally the material composition help designer to detect the special wavelength in terahertz spectrum region.

2.6 Terahertz and Infrared Photodetector Based on Electromagnetically Induced Transparency

The noise source and the dark current inhibit the correct detection of the low-level terahertz signals. A great deal of research has gone into the elimination of the noise source in photodetectors [27, 135]. Until now in practical implementations efforts are usually limited to cooling of the designed devices, which is hard from a practical point of view. The environment temperature controlling is critical and a hard problem especially in room temperature due to high-level dark current. The target radiation detection is limited by the thermionic emission dark current. The photon energy in the terahertz level detection may be in the order of the thermal broadening of electron distribution (KT ~ 6 meV at 77 K and 25 meV at 300 K) [1, 86].

However in the electromagnetically induced transparency (EIT) based photodetection [136–138], the electrons are not directly excited by terahertz radiation. On the other hand the absorption characteristic of a short-wavelength probe optical field is affected by terahertz radiation. In this system, the important thermionic dark current can be nearly cancelled out. In this part, the terahertz detection based on EIT process is studied in a multi-level atomic systems which finally realized by quantum well structures. The mathematical background for these structures is developed and then it is shown that the probe field absorption can be controlled by the electromagnetic control and terahertz radiations.

2.6.1 Electromagnetically Induced Transparency Phenomena

During the past two decades, quantum coherence (atomic phase coherence) effects have exhibited many physically interesting phenomena such as electromagnetically induced transparency (EIT) [139]. The control of linear and non-linear optical properties of a material system using resonant electromagnetic

2.6 Terahertz and Infrared Photodetector

fields has become more and more important in recent years. EIT has proved to be a powerful technique that can be used to eliminate the effect of a medium on a propagating beam of electromagnetic radiation, while retaining the large and desirable nonlinear optical properties associated with the resonant response of a medium [139–141]. Recent remarkable applications include ultraslow light pulse propagation and light storage in atomic vapour systems. The occurrence of EIT in other configurations has also been predicted theoretically. Most of this work has been investigated in the atomic vapour systems, although an extension of these ideas to solid-state systems would be more promising. In the next part the electromagnetically induced transparency are discussed theoretically.

2.6.1.1 Atom–Field Interaction Hamiltonian

An electron with charge e and mass m interacting with an external electromagnetic field is described by a minimal-coupling Hamiltonian as [140]:

$$H = \frac{1}{2}[P - eA(r,t)]^2 + eU(r,t) + V(r) \tag{2.81}$$

where P is the canonical momentum operator, $A(r, t)$ and $U(r, t)$ are the vector and scalar potentials of the external field, respectively and $V(r)$ is an electrostatic potential that is normally the atomic binding potential. In this section, we first derive this Hamiltonian from a gauge invariance point of view, before reducing it to a simple form suitable for describing the interaction of a two-level atom with the radiation field. We examine the problem of an electron bound by a potential $V(r)$ to a force center (nucleus) located at r_0. The minimal-coupling Hamiltonian for an interaction between an atom and the radiation field can be reduced to a simple form by using the dipole approximation. The entire atom is immersed in a plane electromagnetic wave described by a vector potential $A(r_0 + r, t)$:

$$\begin{aligned} A(r_0 + r, t) &= A(t)\exp(\mathrm{i}K \cdot (r_0 + r)) \\ &= A(t)\exp(\mathrm{i}K \cdot r_0)(1 + \mathrm{i}K \cdot r + \cdots) \end{aligned} \tag{2.82}$$

This vector potential may be written in the dipole approximation, $K \cdot r \ll 1$, as:

$$A(r_0 + r, t) \cong A(t)\exp(\mathrm{i}K \cdot r_0) \tag{2.83}$$

The Schrodinger equation for this problem (in the dipole approximation) is given by:

$$\left\{ \frac{\hbar^2}{2m}\left[\nabla - \frac{\mathrm{i}e}{\hbar}A(r_0,t)\right]^2 + V(r) \right\}\psi(r,t) = \mathrm{i}\hbar\frac{\partial}{\partial t}\Psi(r,t) \tag{2.84}$$

where we are working in the radiation gauge, in which:

$$U(r,t) = 0$$
$$\nabla \cdot A = 0 \tag{2.85}$$

We have added the term $V(r)$ in the Hamiltonian which arises from the electrostatic potential that binds the electron to the nucleus. For simplicity in Eq. 2.84 we define a new wave function as:

$$\Psi(r,t) = \exp\left[\frac{ie}{\hbar}A(r_0,t)\right]\Phi(r,t) \tag{2.86}$$

By inserting Eq. 2.85 into Eq. 2.84 we have:

$$i\hbar\dot{\Phi}(r,t) = H\Phi(r,t) \tag{2.87}$$

where

$$H = H_0 + H_1 \tag{2.88}$$

$$H_0 = \frac{p^2}{2m} + V(r) \tag{2.89}$$

$$H_1 = -er \cdot E(r_0 \cdot t) \tag{2.90}$$

In the above equations H_0 is the unpertubated Hamiltonian and H_1 is the pertubated Hamiltonian, respectively.

2.6.1.2 Equation of Motion for the Density Matrix

In many situations we may not know $|\phi\rangle$ only know the probability $P\varphi$ that the system is in the state $|\phi\rangle$. For such a situation, we define the density operator ρ as [140]:

$$\rho = \sum P_\Psi |\Psi\rangle\langle\Psi| \tag{2.91}$$

We obtain the equation of motion for the density matrix from the Schrodinger equation as:

$$|\dot{\varphi}\rangle = -\frac{i}{\hbar}H|\varphi\rangle \tag{2.92}$$

With taking the time derivative of ρ we have:

$$\dot{\rho} = \sum_\varphi P_\varphi(|\dot{\varphi}\rangle\langle\varphi| + |\varphi\rangle\langle\dot{\varphi}| \tag{2.93}$$

where P_φ, is time independent. By using Eqs. 2.92 and 2.93 we have:

2.6 Terahertz and Infrared Photodetector

$$\dot{\rho} = -\frac{i}{\hbar}[H, \rho] \tag{2.94}$$

In the above equation we have not included the decay rates of the atomic levels due to spontaneous emission. The decay rates can be incorporated in the above equation by a relaxation matrix Γ, which is denoted by the equation:

$$\langle n|\Gamma|m\rangle = \gamma_n \delta_{nm} \tag{2.95}$$

With this addition, the density matrix equation of motion becomes:

$$\dot{\rho} = \frac{-i}{\hbar}[H, \rho] - \frac{1}{2}\{\Gamma, \rho\} \tag{2.96}$$

where $\{\Gamma, \rho\} = \Gamma\rho + \rho\Gamma$. In general, the equation of motion becomes as:

$$\dot{\rho}_{ij} = -\frac{i}{\hbar}\sum_k \left(H_{ik}\rho_{Kj} - \rho_{iK}H_{kj}\right) - \frac{1}{2}\sum_k \left(\Gamma_{ik}\rho_{kj} + \rho_{ik}\Gamma_{kj}\right)c \tag{2.97}$$

2.6.1.3 Coherent Trapping: Dark States

It is possible to cancel absorption or emission under certain conditions. This interesting phenomenon in which a coherent superposition of atomic states is responsible for a novel effect is coherent trapping. If an atom is prepared in a coherent superposition of states, it is possible to cancel absorption or emission under certain conditions. So these atoms are then effectively transparent to the incident field even, the presence of resonant transitions. We discuss the effect of coherent trapping and dark state in three-level atomic systems. Figure 2.51 shows the three level atomic systems where the dark states are shown.

The Hamiltonian for the system, in the rotating-wave approximation, is obtained by:

$$H_0 + H_1 \tag{2.98}$$

$$H_0 = \hbar\omega_a|a\rangle\langle a| + \hbar\omega_b|b\rangle\langle b| + \hbar\omega_c|c\rangle\langle c| \tag{2.99}$$

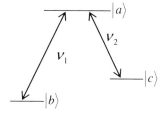

Fig. 2.51 Interaction of three level atomic systems with two single mode fields

$$H_1 = -\frac{\hbar}{2}\left(\Omega_{R1}e^{-i\varphi_1}e^{-iv_1 t}|a\rangle\langle b| + \Omega_{R2}e^{-i\varphi_2}e^{-iv_2 t}|a\rangle\langle c|\right) + \text{H.c.} \quad (2.100)$$

where $\Omega_{R1}e^{-i\varphi_1}$ and $\Omega_{R2}e^{-i\varphi_2}$ are the complex Rabi frequencies associated with the coupling of the field modes of frequencies v_1, v_2 to the atomic transitions $|a\rangle \rightarrow |b\rangle$ and $|a\rangle \rightarrow |c\rangle$. The atomic wave function can be written as:

$$|\Psi\rangle = C_a(t)e^{-i\omega_a t}|a\rangle + C_b(t)e^{-i\omega_b t}|b\rangle + C_c(t)e^{-i\omega_c t}|c\rangle \quad (2.101)$$

The probability amplitudes $C_a(t)$, $C_b(t)$, and $C_c(t)$ can be derived from the Schrödinger equation $i\hbar|\dot\Psi\rangle = H|\Psi\rangle$ as:

$$\dot C_a = \frac{i}{2}\left(\Omega_{R1}e^{-i\Phi_1}C_b + \Omega_{R2}e^{-i\Phi_1}C_c\right) \quad (2.102)$$

$$\dot C_b = \frac{i}{2}\left(\Omega_{R1}e^{-i\Phi_1}C_a\right) \quad (2.103)$$

$$\dot C_c = \frac{i}{2}\left(\Omega_{R2}e^{-i\Phi_2}C_a\right) \quad (2.104)$$

By the analytical solution we can find the probability amplitudes as:

$$C_a(t) = 0$$

$$C_b(t) = \frac{1}{\sqrt{2}}$$

$$C_c(t) = \frac{1}{\sqrt{2}}e^{-i\varphi} \quad (2.105)$$

It is clear that the population is trapped in the lower states and there is no absorption even in the presence of the electromagnetic field. This means that the absorption can be changed by this method. In the present three-level atom, coherent trapping occurs due to the destructive quantum interference between the two transitions. Finally the wave function for this system is obtained as:

$$|\Psi(t)\rangle = \frac{\Omega_{R2}(t)e^{-i\varphi_2}|b\rangle - \Omega_{R1}(t)e^{-i\varphi_1}|c\rangle}{\sqrt{\Omega_{R1}^2 + \Omega_{R2}^2}} \quad (2.106)$$

We discussed the phenomena of coherent population trapping in three-level atom system in which the lower levels are prepared in a coherent superposition state. The other related phenomenon is the EIT. When a pair of near-resonant laser fields interact with the three levels (lambda type), in the condition of two-photon resonance, the populations are coherently trapped in the two lower levels. At this two-photon resonance point, the atom is in a dark state and decoupled with the

2.6 Terahertz and Infrared Photodetector

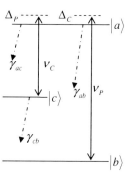

Fig. 2.52 The EIT phenomena in three-level atomic systems

applied fields, so the fields transmit the medium without absorption and the EIT occurs. Figure 2.52 shows the three-level atomic system which EIT phenomena are happens.

Due to no absorption at the transparency point, EIT is applied to various low light nonlinear optical processes. The linear and nonlinear susceptibilities have been resonantly enhanced because of the coherent control of quantum states. Based on EIT, various kinds of four-wave and six-wave mixing with high quantum efficiency are investigated.

The dipole moment in quantum mechanics is defined by $P = e\langle\Psi(t)|r|\Psi(t)\rangle$. In this case we obtain dipole moment as [1, 27]:

$$P = \rho_{ab}\wp_{ba} + \rho_{ac}\wp_{ca} + \rho_{cb}\wp_{cb} + \text{c.c.} \quad (2.107)$$

where $\wp_{ij} = \langle i|r|j\rangle$ is the matrix element. In the selection rules theory the transition $|c\rangle \rightarrow |b\rangle$ is not valid, so $\wp_{bc} = \wp_{cb} = 0$. The probe frequency and the transition $|a\rangle \rightarrow |c\rangle$ frequency are different ($\rho_{ab} \rightarrow 0$) so the dipole moment can be written as:

$$P(v) = \rho_{ab}\wp_{ba} + \text{c.c.} \quad (2.108)$$

On the other hand we can write:

$$P(z,t) = \varepsilon_0 \int_0^\infty d\tau \tilde{x}(\tau) E(z, t-\tau) \quad (2.109)$$

where E is flat wave across the z axis, therefore:

$$E(z,t) = 1/2 E e^{-i(vt-kz)} + \text{c.c.} \quad (2.110)$$

Finally we obtain the dipole moment as:

$$P(z,t) = \frac{\varepsilon_0 E}{2} \int_0^\infty d\tau \tilde{\chi}(\tau) \left(e^{-i(v(t-\tau)-kz)} + e^{i(v(t-\tau)-kz)} \right)$$

$$= \frac{\varepsilon_0 E}{2} \left(\left(\int_0^\infty \tilde{\chi}(\tau)(e^{iv\tau} d\tau) \right) e^{-i(vt-kz)} + \left(\int_0^\infty \tilde{\chi}(\tau) \cdot e^{-iv\tau} d\tau \right) e^{i(vt-kz)} \right)$$

$$= \frac{\varepsilon_0 E}{2} \left(\chi(-v) e^{-i(vt-kz)} + \chi(v) e^{i(vt-kz)} \right)$$

$$= \frac{\varepsilon_0 E}{2} \chi(v) e^{-i(vt-kz)} + \text{c.c.}$$

(2.111)

By using Eqs. 2.107 and 2.110 we obtain:

$$\frac{\varepsilon_0 E}{2} \chi(v) e^{-i(vt-kz)} + \text{c.c.} = \rho_{ab} \wp_{ba} + \text{c.c.} \qquad (2.112)$$

In a system with atom density N_a we write:

$$\frac{\varepsilon_0 E}{2} \chi(v) e^{ivt} = (\rho_{ab} \wp_{ba})^* N_a \qquad (2.113)$$

And finally the susceptibility is equal:

$$\chi(v) = \frac{2}{\varepsilon_0 E} N_a (\rho_{ab} \wp_{ba})^* e^{-ivt} \qquad (2.114)$$

For solving above equation we should find the ρ_{ab} from density matrix, so in the three-level atomic system the total Hamiltonian and density matrix can obtain as [136, 140]:

$$H = \hbar \begin{bmatrix} v_b & 0 & -\frac{1}{2}\Omega_p e^{-iv_p t} \\ 0 & v_c & -\frac{1}{2}\Omega_c e^{-iv_p t} \\ -\frac{1}{2}\Omega_p e^{-iv_p t} & -\frac{1}{2}\Omega_c e^{-iv_p t} & v_a \end{bmatrix} \qquad (2.115)$$

$$\dot{\rho}_{ab} = -(iv_{ab} + \gamma_{ab})\rho_{ab} - \frac{i}{2}\Omega_p e^{-iv_p t}(\rho_{aa} - \rho_{bb}) + \frac{i}{2}\Omega_c e^{-iv_c t}\rho_{cb} \qquad (2.116)$$

$$\dot{\rho}_{cb} = -(iv_{cb} + \gamma_{cb})\rho_{cb} - \frac{i}{2}\Omega_p e^{-iv_p t}\rho_{ca} + \frac{i}{2}\Omega_c e^{iv_c t}\rho_{ab} \qquad (2.117)$$

$$\dot{\rho}_{ac} = -(iv_{ac} + \gamma_{ac})\rho_{ac} - i\Omega_c e^{-iv_c t}(\rho_{aa} - \rho_{cc}) + \frac{i}{2}\Omega_p e^{-iv_p t}\rho_{bc} \qquad (2.118)$$

With the using Eq. 2.113 we obtain optical susceptibility as:

2.6 Terahertz and Infrared Photodetector

$$\chi^{(1)}(v) = \frac{iN_a\wp^2\left(i[\Delta_p - \Delta_c] + \gamma_{cb}\right)}{\varepsilon_0\hbar\left[(i\Delta_p + \gamma_{ab})(i[\Delta_p - \Delta_c] + \gamma_{cb}) + \Omega_c^2/4\right]} \quad (2.119)$$

The linear response of atom for resonant light determined by the first order of susceptibility as [140]:

$$\chi^{(1)} = \chi' + i\chi'' \quad (2.120)$$

The real and imaginary parts of the susceptibility are proportional to absorption coefficient and refractive coefficient, respectively, so we can write:

$$\alpha = \omega_p n_0 \chi''/c \quad (2.121)$$

$$\beta = \omega_p n_0 \chi'/2c \quad (2.122)$$

Imaginary and real parts of susceptibility for interaction of three level atomic system with two single mode fields are shown in Fig. 2.53. In the case when to field are resonant because of the dark state and coherent trapping and related EIT phenomena the absorption and refractive characteristic are modified. It is illustrated that we can change the absorption and transmission coefficient of electromagnetic field by another electromagnetic field names control field.

2.6.2 EIT-Based Photodetection

The terahertz photodetection system based on EIT phenomena is schematically shown in Fig. 2.54. This system involves a 4-level atomic system interacting with three fields of control, probe and terahertz electromagnetic signals [136, 138].

The total Hamiltonian in the rotating-wave approximation (keeping only energy conserving terms in Hamiltonian) and in the ignored counter rotating terms case [140], is given as follows:

$$H = H_0 + H_1 \quad (2.123)$$

where

$$H_0 = \hbar\omega_1|1\rangle\langle 1| + \hbar\omega_2|2\rangle\langle 2| + \hbar\omega_3|3\rangle\langle 3| + \hbar\omega_4|4\rangle\langle 4| \quad (2.124)$$

$$H_1 = \hbar\Omega_p|1\rangle\langle 4| + \hbar\Omega_c|2\rangle\langle 4| + \hbar\Omega_{IR1}|2\rangle\langle 3| \quad (2.125)$$

Here Ω_p, Ω_c and Ω_{IR1} are the Rabi frequencies associated with the coupling of the field modes of, probe, control and terahertz signals to the atomic transition states 2–4, 4–1 and 3–2, respectively. If some mathematical rearranging and

Fig. 2.53 **a** Imaginary and **b** real parts of susceptibility for interaction of three level atomic system with two single mode fields

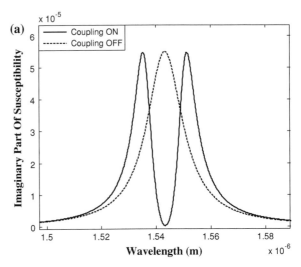

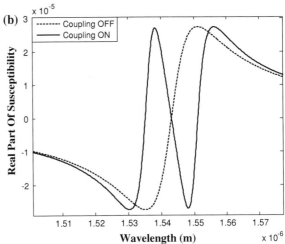

Fig. 2.54 Schematic of terahertz photodetection system based on EIT [136, 138]

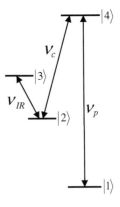

2.6 Terahertz and Infrared Photodetector

manipulation is done, we can obtain the following analytical density matrix time development equations which handle this 4-level atomic system [136, 138, 140]:

$$\dot{\rho}_{41} = -[i\Delta_p + \gamma_{41}]\rho_{41} + i\Omega_p\rho_{11} + i\Omega_c\rho_{21} - i\Omega_p\rho_{44} \qquad (2.126)$$

$$\dot{\rho}_{21} = -[i(\Delta_p - \Delta_c) + \gamma_{21}]\rho_{21} + i\Omega_{IR}\rho_{31} + i\Omega_c\rho_{41} - i\Omega_p\rho_{24} \qquad (2.127)$$

$$\dot{\rho}_{31} = -[i(\Delta_p + \Delta_{IR} - \Delta_c) + \gamma_{31}]\rho_{31} + i\Omega_{IR}\rho_{21} - i\Omega_p\rho_{34} \qquad (2.128)$$

where $\Delta_p = v_{41} - v_p$, $\Delta_c = v_{42} - v_c$ and $\Delta_{IR} = v_{32} - v_{IR}$ are detuning corresponding to probe, control and terahertz signals, respectively. The differential matrix equation can be solved as follow:

$$\dot{R} = -MR + A \qquad (2.129)$$

where $A = \begin{bmatrix} i\alpha \\ 0 \\ 0 \end{bmatrix}$ and

$$M = \begin{bmatrix} (i\Delta_p + \gamma_{41}) & -i\Omega_c & 0 \\ -i\Omega_c & [i(\Delta_p - \Delta_c) + \gamma_{21}] & -i\Omega_{IR} \\ 0 & -i\Omega_{IR} & [i(\Delta_p + \Delta_{IR} - \Delta_c) + \gamma_{31}] \end{bmatrix}.$$

Also, γ_{41}, γ_{21}, γ_{31}, Ω_c and Ω_{IR} are decay rate of density matrix equations, Rabi frequency of infrared and control field, respectively, and $R = \begin{bmatrix} \rho_{41} \\ \rho_{21} \\ \rho_{31} \end{bmatrix}$. The exact analytical solution is:

$$R(t) = M^{-1}A \qquad (2.130)$$

The optical susceptibility of probe field is obtained by:

$$\chi_{41} = \frac{2N_a\wp^2/\varepsilon_0\hbar\left[\gamma_{21} + i(\Delta_p - \Delta_c) + \frac{\Omega_{IR}^2}{\alpha}\right]}{\left[\Omega_c^2 + [i\Delta_p + \gamma_{41}][i(\Delta_p - \Delta_c) + \gamma_{21}] + \frac{\Omega_{IR}^2(i\Delta_p + \gamma_{41})}{\alpha}\right]} \qquad (2.131)$$

where Ω_p and N_a are the Rabi frequency of the probe field and the atomic density, respectively. One should note that the transmission coefficient is proportional to the imaginary part of the calculated susceptibility. In our calculation, the environment temperature effect on the system operation is added by $E_T = \frac{3}{2}KT$ where E_T, K and T are the thermal energy, the Boltzmann constant and temperature in Kelvin, respectively. This energy should be added to the signal energy. On the other hand, the exact behavior of 4-level atomic system can be determined by calculating zeros and poles of matrix M or calculating zeros and poles of ρ_{41}. In the case of without detuning, that is all fields are coupled, we have:

$$\text{poles}(\rho_{41}) = \begin{cases} v_{41} \\ v_{41} + \sqrt{\Omega_c^2 + \Omega_{IR}^2} \\ v_{41} - \sqrt{\Omega_c^2 + \Omega_{IR}^2} \end{cases} \quad \text{and} \quad \text{zeros}(\rho_{41}) = \begin{cases} v_{41} + \Omega_{IR} \\ v_{41} - \Omega_{IR} \end{cases} \quad (2.132)$$

$$\text{If } \Omega_{IR=0} \Rightarrow \text{poles}(\rho_{41}) = \begin{cases} v_{41} \\ v_{41} + \Omega_c^2 \\ v_{41} - \Omega_c^2 \end{cases} \quad \text{and} \quad \text{zeros}(\rho_{41}) = \begin{cases} v_{41} \\ v_{41} \end{cases} \quad (2.133)$$

However in the presence of detuning, the locations of zeros and poles change and the response of the system is completely different. The important case is, when $\Delta_{IR} = \Delta_c$ we have the following result:

$$\text{poles}(\rho_{41}) = \begin{cases} v_{41} \\ v_{41} + \sqrt{\Omega_c^2 + \Omega_{IR}^2} \\ v_{41} - \sqrt{\Omega_c^2 + \Omega_{IR}^2} \end{cases} \quad \text{and} \quad \text{zeros}(\rho_{41}) = \begin{cases} v_{41} + \Omega_{IR} \\ v_{41} - \Omega_{IR} \end{cases} \quad (2.134)$$

In this section, simulated results including the optical susceptibilities and transmission coefficient in the different proposed atomic systems are presented and discussed. First, we illustrate the effects of infrared and control fields on the real and imaginary parts of the optical susceptibility for 4-level structure (Fig. 2.54) in Fig. 2.55a, b. In Fig. 2.55a, b, it is shown that for a given control field, when the magnitude of infrared field increases the spectrum of the imaginary and real parts of the optical susceptibility are displaced and changed. In the presence of control field, an Autler–Townes doublet is created and when the infrared field is applied, Autler–Townes doublet broadens and another peak appears in the central part of the spectrum of doublet, so three peaks are appeared in the susceptibility spectra. This broadness and central peak increase with increasing of the infrared field intensity. This behavior was expected from the calculation of poles and zeros (Eq. 2.132). As we know, this central peak disappears, when the infrared field is switched off.

It is clear that, the transmission and reflection coefficients are proportional to imaginary and real parts of the susceptibility, respectively. The effect of infrared and control fields on the transmission coefficient of the probe field for 4-level system is shown in Fig. 2.56. It is observed that the Autler–Townes doublet is created with the definite control field and is broadened with increasing of the control field. In the presence of IR field, central absorption peak appears. Thus, measuring of the optical wavelength absorption illustrates the infrared signal level.

Figure 2.57 shows the effect of different IR intensity on the transmission coefficient spectrum. It is observed that with increasing IR intensity the transmission coefficient at central peak also decreases. When IR intensity becomes comparable with control field the width of Autler–Townes doublet increases and is considerable. This behavior was expected from the calculated poles and zeros (Eq. 2.132). It is clear that with changing the IR field intensity, position of the poles and zeros are changed and so the splitting of doublet changes also.

2.6 Terahertz and Infrared Photodetector

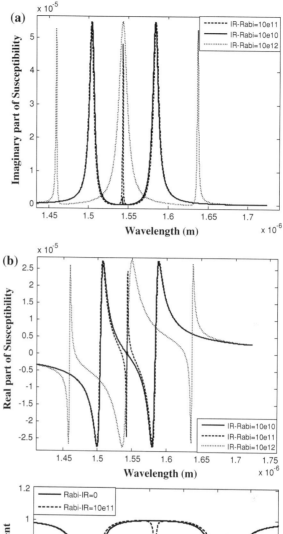

Fig. 2.55 **a** Real and **b** imaginary parts of susceptibility versus wavelength for different Infrared Rabi frequencies ($\Omega_c = 5 \times 10^{12}\,\text{s}^{-1}$, $\Omega_p = 10^8\,\text{s}^{-1}$, $\Delta_{IR} = \Delta_c = 0$, $N_a = 10^{20}\,\text{cm}^{-3}$, $\gamma_{41} = 10^{12}\,\text{s}^{-1}$, $\gamma_{31} = \gamma_{21} = 5 \times 10^9\,\text{s}^{-1}$, $\wp = 10^{-10}\,\text{e} - \text{cm}$)

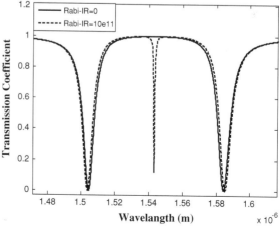

Fig. 2.56 Transmission coefficient versus wavelength with and without IR field [136] ($\Omega_c = 5 \times 10^{12}\,\text{s}^{-1}$, $\Omega_p = 10^8\,\text{s}^{-1}$, $\Delta_{IR} = \Delta_c = 0$, $N_a = 10^{20}\,\text{cm}^{-3}$, $\gamma_{41} = 10^{12}\,\text{s}^{-1}$, $\gamma_{31} = \gamma_{21} = 5 \times 10^9\,\text{s}^{-1}$, $\wp = 10^{-10}\,\text{e} - \text{cm}$)

Fig. 2.57 Transmission coefficient versus wavelength with different infrared Rabi frequencies [136] ($\Omega_c = 5 \times 10^{12}\text{s}^{-1}, \Omega_p = 10^8\text{s}^{-1}, \Delta_{IR} = \Delta_c = 0, N_a = 10^{20}\text{cm}^{-3}, \gamma_{41} = 10^{12}\text{s}^{-1}, \gamma_{31} = \gamma_{21} = 5 \times 10^9\text{s}^{-1}, \wp = 10^{-10}\text{e}-\text{cm}$)

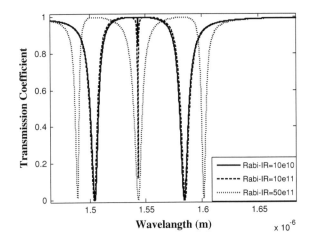

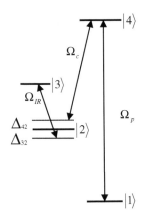

Fig. 2.58 Schematic of 4-levels atomic system with detuned fields [136, 138]

In previous simulations, we assumed that all fields are on resonance and there is not any detuning. However if we consider nonzero detuning for applied electric fields (Fig. 2.54), so, we have the following scheme illustrated in Fig. 2.58.

In this case the position of poles and zeros are changed and so the imaginary and real parts of the susceptibility as well as the transmission coefficient should be changed. The appeared central absorption peak shifts left or right according to positive or negative detuning (Fig. 2.59). It is interesting when $\Delta_{IR} = \Delta_c$, the central absorption peak is fixed and does not change (Fig. 2.59). Therefore, with control of the control field detuning effect of IR signal level on optical absorption peak displacement can be ignored.

Figure 2.60 shows two different situations. When all three fields are on resonance ($\Delta_{IR} = 0$), there is an absorption exactly in central part of the transmission band (dashed line). If the IR field is a long-wavelengths infrared field and the probe field an optical field, so, the absorption characteristic of the short-wavelength probe field on resonance translate the absorption characteristic of IR field. These features make

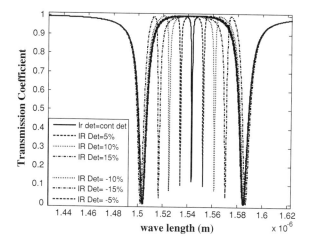

Fig. 2.59 Transmission coefficient versus wavelength with different detuning [136] ($\Omega_c = 5 \times 10^{12}\,\text{s}^{-1}, \Omega_p = 10^8\,\text{s}^{-1}, \Delta_{IR} = \Delta_c = 0, N_a = 10^{20}\,\text{cm}^{-3}, \gamma_{41} = 10^{12}\,\text{s}^{-1}, \gamma_{31} = \gamma_{21} = 5 \times 10^9\,\text{s}^{-1}, \wp = 10^{-10}\,\text{e} - \text{cm}$)

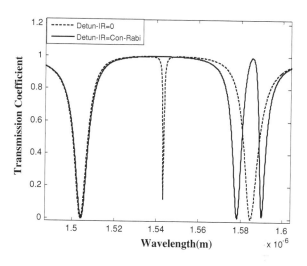

Fig. 2.60 Transmission coefficient versus wavelength without and with Infrared detuning equal with Rabi frequency of control field [136] ($\Omega_c = 5 \times 10^{12}\,\text{s}^{-1}, \Omega_p = 10^8\,\text{s}^{-1}, \Delta_{IR} = \Delta_c = 0, N_a = 10^{20}\,\text{cm}^{-3}, \gamma_{41} = 10^{12}\,\text{s}^{-1}, \gamma_{31} = \gamma_{21} = 5 \times 10^9\,\text{s}^{-1}, \wp = 10^{-10}\,\text{e} - \text{cm}$)

the present system suitable for a kind of detector. If IR field detuning becomes $\Delta_{IR} \approx \Omega_c$ and other fields are on resonance a new transparency appears near the maximum absorption of the system. Thus it is also possible to turn absorption into transparency.

Figure 2.61 shows the effect of environment temperature on the transparency spectrum of the probe field. The effect of the environment temperature is the same as detuning effect. It is shown that wavelength of the central absorption changes with temperature. It is observed that 7 nm shift of wavelength for 30°C of temperature fluctuation. When refractive index of material increases ($n_{\text{refractive}} = 1 \rightarrow 3.5\,(\text{for GaAs})$), the wavelength temperature coefficient decreases ($\Delta\lambda/\Delta T = 0.23\,\text{nm}/°\text{C} \rightarrow 66\,\text{pm}/°\text{C}$) that is so excellent from practical implementation point of view.

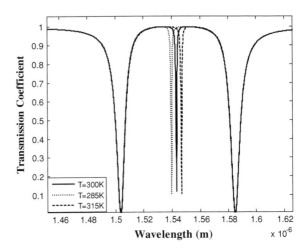

Fig. 2.61 Transmission coefficient versus wavelength with different environment temperatures [136] ($\Omega_c = 5 \times 10^{12}$ s^{-1}, $\Omega_p = 10^8$ s^{-1}, $\Delta_{IR} = \Delta_c = 0$, $N_a = 10^{20}$ cm^{-3}, $\gamma_{41} = 10^{12}$ s^{-1}, $\gamma_{31} = \gamma_{21} = 5 \times 10^9$ s^{-1}, $\wp = 10^{-10}$ e − cm)

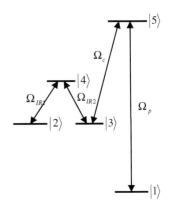

Fig. 2.62 Schematic of 5-levels atomic system [136]

In this part, we investigate interaction of IR, control and probe signals in the case of 5-level atomic system which is proposed in Fig. 2.62. Figure 2.63 shows the effect of infrared radiation on 5-level atomic system. In this system two equal infrared signals are used. In this system control field introduces the Autler–Townes doublet in the real and imaginary parts of the susceptibility as well as the splitting of doublet is proportional to the control Rabi frequency. In the presence of IR field, two new doublets are observed in the band gap. In the next part, we illustrate exact result in some schemes. For managing of thermal problems in detectors, we propose the schematic of 5-level atomic system. In this system the effect of environment temperature is added to the incoming energy of photon and this is like having detuned in the fields. If we consider $\Delta_{IR} = \Delta_c$ in Fig. 2.62, we can cancel out this temperature dependency. In this system, we use two equal states for the incoming IR radiation. This means that the thermal energy that added to the two IR radiations and the detuning of two IR is the same.

2.6 Terahertz and Infrared Photodetector

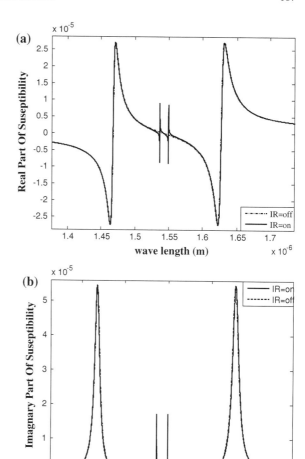

Fig. 2.63 **a** Real and **b** imaginary parts of susceptibility versus wavelength with infrared radiation on (*solid line*) and infrared radiation off (*dashed line*) [136] $\Omega_c = 10^{12} s^{-1}$, $\Omega_p = 10^8 s^{-1}$, $\Omega_{IR} = 5^{11}$, $\Delta_{IR1} = \Delta_{IR2} = \Delta_c = \Delta_p = 0$, $T = 0\,K$, $N_a = 10^{20} cm^{-3}$, $\gamma_{51} = 10^{12} s^{-1}$, $\gamma_{31} = \gamma_{21} = 5 \times 10^9 s^{-1}$, $\gamma_{41} = 10^{11}$, $\wp = 10^{-10} e - cm$

As disused in the above the optical susceptibility in a five-level atomic system for the probe field can be obtain as:

$$\dot{\rho}_{51} = -[(i\Delta_p) + \gamma_{51}]\rho_{51} + i\Omega_c\rho_{31} + i\Omega_p\rho_{11} - i\Omega_p\rho_{55} \quad (2.135)$$

$$\dot{\rho}_{31} = -[i(\Delta_p - \Delta_c) + \gamma_{31}]\rho_{31} + i\Omega_c\rho_{51} + i\Omega_{IR2}\rho_{41} - i\Omega_p\rho_{35} \quad (2.136)$$

$$\dot{\rho}_{41} = -[i(\Delta_p - \Delta_c + \Delta_{IR2}) + \gamma_{41}]\rho_{41} + i\Omega_{IR2}\rho_{31} + i\Omega_{IR1}\rho_{21} - i\Omega_p\rho_{45} \quad (2.137)$$

$$\dot{\rho}_{21} = -[i(\Delta_p - \Delta_c + \Delta_{IR2} - \Delta_{IR1}) + \gamma_{21}]\rho_{21} + i\Omega_{IR1}\rho_{41} - i\Omega_p\rho_{25} \quad (2.138)$$

$$\chi_P^{(1)} = \frac{i2N\wp^2/\varepsilon_0\hbar \left[L_2 L_3 + \Omega_1^2 + \frac{L_3\Omega_2^2}{L_1}\right]}{\left[L_4 L_3 L_2 + L_4\Omega_1^2 + L_2\Omega_c^2 + \frac{L_3 L_4 \Omega_2^2}{L_1}\right]} \quad (2.139)$$

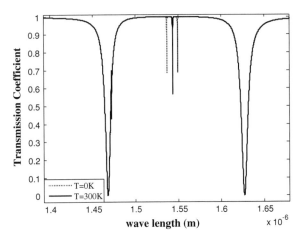

Fig. 2.64 Transmission coefficient versus wavelength with different environment temperature [136] $\Omega_c = 10^{12}$ s^{-1}, $\Omega_p = 10^8$ s^{-1}, $\Omega_{IR} = 5^{11}$, $\Delta_{IR1} = \Delta_{IR2} = \Delta_c = \Delta_p = 0$, $T = 0$ K, $N_a = 10^{20}$ cm^{-3}, $\gamma_{51} = 10^{12}$ s^{-1}, $\gamma_{31} = \gamma_{21} = 5 \times 10^9$ s^{-1}, $\gamma_{41} = 10^{11}$, $\wp = 10^{-10}$ e − cm

where $L_1 = [i(\Delta_p - \Delta_c + \Delta_1 - \Delta_2) + \gamma_{ba}]$, $L_2 = [i(\Delta_p - \Delta_c + \Delta_1) + \gamma_{da}]$, $L_3 = [i(\Delta_p - \Delta_c) + \gamma_{ca}]$ and $L_4 = [i(\Delta_p) + \gamma_{ea}]$. In this system, we show that the improper effect of environment temperature on signal energy is reduced. Finally, we cancel out the effect of environment temperature by introducing another atomic system.

Figure 2.64 shows the effect of environment temperature on the operation of this system. The environment temperature energy is added to both of IR radiations. If two infrared radiations are selected to be same therefore effect of thermal energy can be eliminated. It is observed that with increasing the temperature from $T = 0$ K (dashed line) to $T = 300$ K (solid line), one of the central doublet shifts more but the other one shifts as little as 4 nm. If in this case we try to increase the refractive index of mater from 1 to 1.35 [$n_{\text{refractive}} = 1 \rightarrow 3.5$ (for GaAs)], the wavelength shift reduces to 1.1 nm.

Simulated results for the 5-level atomic system presented that in spite of some limitations, this system shows some promise in canceling out the effect environment temperature. In the above atomic systems, we minimized the environment temperature effect on the operation of the systems and cannot cancel out exactly the environment temperature effect.

Figure 2.65a shows the schematic of 6-level atomic system. In this system, we have two control fields and two IR fields, which both IR fields and control fields are the same. We should isolate the environment thermal energy from incoming infrared radiation. The effect of environment temperature is like a detuning in IR field. If we use two IR fields, we can cancel the effect of detuning, because their thermal energies are the same. This 6-levels system may be realized with an asymmetric multi-quantum well structure such as Fig. 2.65b, c.

According to discussed problems the optical susceptibility in 6-levels atomic system, can be written by [136, 138, 140]:

$$\dot{\rho}_{61} = -[(i\Delta_p) + \gamma_{61}]\rho_{61} + i\Omega_{c1}\rho_{31} + i\Omega_{c2}\rho_{41} + i\Omega_p\rho_{11} - i\Omega_p\rho_{66} \quad (2.140)$$

2.6 Terahertz and Infrared Photodetector

Fig. 2.65 a Schematic of 6-level atomic system, b realization of 6-level atomic system with quantum well, and c the effect of coupling well on splitting of energy levels [136]

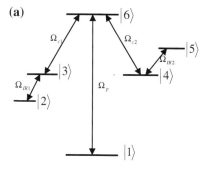

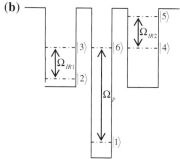

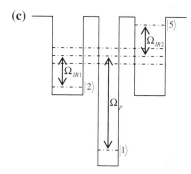

$$\dot{\rho}_{31} = -\left[i(\Delta_p - \Delta_{c1}) + \gamma_{31}\right]\rho_{31} + i\Omega_{c1}\rho_{61} + i\Omega_{IR1}\rho_{21} - i\Omega_p\rho_{36} \quad (2.141)$$

$$\dot{\rho}_{41} = -\left[i(\Delta_p - \Delta_{c2}) + \gamma_{41}\right]\rho_{41} + i\Omega_{c2}\rho_{61} + i\Omega_{IR2}\rho_{51} - i\Omega_p\rho_{46} \quad (2.142)$$

$$\dot{\rho}_{51} = -\left[i(\Delta_p - \Delta_{c2}) + \gamma_{51}\right]\rho_{51} + i\Omega_{IR2}\rho_{41} - i\Omega_p\rho_{56} \quad (2.143)$$

$$\dot{\rho}_{21} = -\left[i(\Delta_p - \Delta_{c1} - \Delta_{IR1}) + \gamma_{21}\right]\rho_{21} + i\Omega_{IR1}\rho_{31} - i\Omega_p\rho_{26} \quad (2.144)$$

$$\chi_{41}^{(1)} = \left[\frac{2N\wp^2/\varepsilon_0\hbar\left[-L_4L_5 + \frac{-\Omega_{IR1}^2[L_5]}{L_1} + \frac{-\Omega_{IR2}^2[L_4]}{L_2} + \frac{-\Omega_{IR1}^2\Omega_{IR2}^2}{L_1L_2}\right]}{\left[\frac{-L_3\Omega_{IR1}^2\Omega_{IR2}^2}{L_1L_2}\right] + \left[\frac{-L_3L_4\Omega_{IR2}^2}{L_2}\right] + \left[\frac{-L_3L_5\Omega_{IR1}^2}{L_1}\right] + [-L_3L_4L_5] + [-L_4\Omega_{c2}^2] + [-L_5\Omega_{c1}^2] + \left[\frac{-\Omega_{c1}^2\Omega_{IR2}^2}{L_2}\right] + \left[\frac{-\Omega_{c2}^2\Omega_{IR1}^2}{L_1}\right]}\right]$$

$$(2.145)$$

where $L_1 = [i(\Delta_p - \Delta_{c1} - \Delta_{IR1}) + \gamma_{21}]$, $L_2 = [i(\Delta_p + \Delta_{c1} + \Delta_{IR2}) + \gamma_{61}]$, $L_3 = [i\Delta_p + \gamma_{41}]$, $L_4 = [i(\Delta_p - \Delta_{c1}) + \gamma_{31}]$ and $L_5 = [i(\Delta_p + \Delta_{c2}) + \gamma_{51}]$.

We show that the operation of the photodetector design based on this atomic system is completely independent from the environment temperature.

Now, we show that using 6-level atomic system the effect of environment temperature completely can be removed. This behavior of the new structure is very interesting from the room temperature terahertz detectors point of view. Figure 2.66 shows the effect of these two IR signals on the optical susceptibility. If one of the control fields is on, Autler–Townes doublet is generated and when the other control field becomes also active, Autler–Townes splits more. In the presence of IR fields, we see a small broadness of Autler–Townes doublet. Also it is observed that a sharp impulse in the band gap generated which magnitude of it changes with the Rabi frequency of infrared signal.

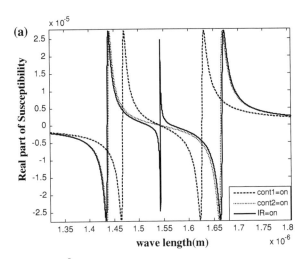

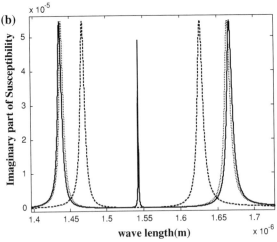

Fig. 2.66 **a** Real and **b** imaginary parts of susceptibility versus wavelength where *dashed line curve* corresponds one of the control fields is on, for dot line both control fields are on and for *solid line*, both control fields and IR fields are on [136] $\Omega_c = 10^{13}$ s^{-1}, $\Omega_p = 10^8$ s^{-1}, $\Omega_{IR} = 5 \times 10^{11}$, $\Delta_{IR1} = \Delta_{IR2} = \Delta_c = \Delta_p = 0$, $T = 0$ K, $N_a = 10^{20}$ cm^{-3}, $\gamma_{61} = 10^{12}$ s^{-1}, $\gamma_{21} = 5 \times 10^9$ s^{-1}, $\gamma_{31} = \gamma_{41} = 10^9$ s^{-1}, $\gamma_{51} = 10^{11}$ s^{-1}, $\wp = 10^{-10}$ e − cm

2.6 Terahertz and Infrared Photodetector

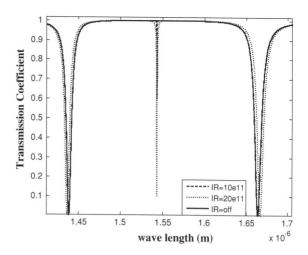

Fig. 2.67 Transmission coefficient versus wavelength with different infrared radiations [136] $\Omega_c = 10^{12}$ s^{-1}, $\Omega_p = 10^8$ s^{-1}, $\Omega_{IR} = 5 \times 10^{11}$, $\Delta_{IR1} = \Delta_{IR2} = \Delta_c = \Delta_p = 0$, $T = 0$ K, $N_a = 10^{20}$ cm^{-3}, $\gamma_{61} = 10^{12}$ s^{-1}, $\gamma_{21} = 5 \times 10^9$ s^{-1}, $\gamma_{31} = \gamma_{41} = 10^9$ s^{-1}, $\gamma_{51} = 10^{11}$ s^{-1}, $\wp = 10^{-10}$ e − cm

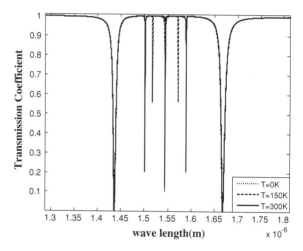

Fig. 2.68 Transmission coefficient versus wavelength with different temperatures [136] $\Omega_c = 10^{12}$ s^{-1}, $\Omega_p = 10^8$ s^{-1}, $\Omega_{IR} = 5^{11}$, $\Delta_{IR1} = \Delta_{IR2} = \Delta_c = \Delta_p = 0$, $N_a = 10^{20}$ cm^{-3}, $\gamma_{61} = 10^{12}$ s^{-1}, $\gamma_{21} = 5 \times 10^9$ s^{-1}, $\gamma_{31} = \gamma_{41} = 10^9$ s^{-1} $\gamma_{51} = 10^{11}$ s^{-1}, $\wp = 10^{-10}$ e − cm

Figure 2.67 shows the effect of different infrared radiations on this atomic system. In this system in the presence of IR signal, it is observed that a sharp impulse generated between the Autler–Townes doublets which magnitude changes with IR radiation. So really IR-photodetector operation is obtained.

Figure 2.68 shows the effect of different environment temperatures on the system operation. It is observed that in presence of temperature two other absorption peaks are appeared and with increasing the applied temperature the observed peaks are broadened. Also, we found that the appeared peak in the central part of the spectrum is independent of temperature. This is interesting for room temperature operation.

In the above section some of simulated results for IR photodetectors based on EIT process were proposed and it was observed that room temperature operation can be concluded. We think that the proposed structures can be realized by quantum wells and dots.

In the conventional infrared photodetectors, the incoming photon or the environment temperature directly excites the ground state electrons to higher energy levels and in the presence of electric field, these electrons are collected as a photocurrent and thermionic dark current [86, 136, 138]. Therefore in high temperature and terahertz applications, the dark current inhibits the correct detection of signal [136, 138]. However in our proposed EIT-based photodetector, the electromagnetic field of terahertz infrared radiation interfere with the electromagnetic field of short-wavelength (1–2 μm) probe field and modify the absorption characteristic of probe field. Therefore the incoming terahertz IR light and the environment temperature do not directly excite electrons, but affect the absorption characteristics of short-wavelength probe optical field. In fact we convert the incoming terahertz IR signal to short-wavelength optical field through EIT phenomena [138, 140], where such problems are not critical in this rang of detection.

In many aspects the eigen-states in quantum well are like atomic systems [140, 141]. In the atomic systems if two fields are coupled and interfere with each other and one of them be strong, so the optical stark effect is created. There is a shift in absorption characteristic, so we see two new absorptions where there was transparency. In this case we say the atomic states are splitting and this is known as an EIT. However in quantum well structures the stark effect can be created with the coupling two wells and the EIT like condition may be appeared [140, 141].

Figure 2.69 shows a 4-subbands asymmetric double quantum well structure. In Fig. 2.69a an infrared field is applied between states 2, 3 and a probe field is applied between states 1, 4. The barrier potential between two well is too thick that two wells and their wave functions are separated (un-coupled). In other word all states are localized. In this case there is no coupling between infrared field and probe field and therefore the EIT like condition is not happen. Thus we are not able to control the absorption characteristic through infrared field. In Fig. 2.69b the

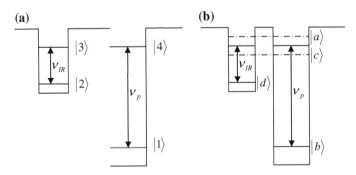

Fig. 2.69 **a** Asymmetric double quantum well with too thick barrier (uncoupled wells) and **b** asymmetric double quantum well with thin barrier (coupled wells) [136]

2.6 Terahertz and Infrared Photodetector

barrier potential (V_0) between two well is thin (coupled wells) and the eigen-states be in the same energy or in the range of electron–LO phonon scattering or electron–electron scattering, so the wave functions of two wells can see each other through thin barrier (resonant-mode). These two states ($|4\rangle, |3\rangle$) are combined and split by $\langle 3|V_0|4\rangle$, so the two new states ($|a\rangle, |c\rangle$) are created. This effect is like to stark effect in an atomic system which is introduced by strong pump field. In the following we show the EIT like pattern can be observed in this quantum well structure.

Using the density matrix formalism, we begin to describe the dynamic response of the proposed 4-subband quantum well structure. By adopting the standard approach (this method has described the results in several experimental papers [142–144] and has been used in several theoretical papers [145–147]) under the electro-dipole and rotating-wave approximations, we can easily obtain the time-dependent density-matrix equations of motion as follows:

$$H = \hbar \begin{bmatrix} v_b & 0 & -\Omega_p e^{iv_p t} & -q\Omega_p e^{iv_p t} \\ 0 & v_d & -\Omega_{IR} e^{iv_{IR} t} & -k\Omega_{IR} e^{iv_{IR} t} \\ -\Omega_p e^{-iv_p t} & -\Omega_{IR} e^{-iv_{IR} t} & v_c & 0 \\ -q\Omega_p e^{-iv_p t} & -k\Omega_{IR} e^{-iv_{IR} t} & 0 & v_4 \end{bmatrix} \quad (2.146)$$

$$\dot{\rho}_{ab} = -i\left(\Delta_p + \frac{\omega_S}{2} + \Gamma_{ab}\right)\rho_{ab} + iq\Omega_p \rho_{bb} - iq\Omega_p \rho_{aa} \\ + ik\Omega_{IR}\rho_{db} - i\Omega_p \rho_{ac} \quad (2.147)$$

$$\dot{\rho}_{db} = -i(\Delta_p - \Delta_{IR} + \Gamma_{db})\rho_{db} + i\Omega_{IR}\rho_{cb} + ik\Omega_{IR}\rho_{ab} \\ - i\Omega_p \rho_{dc} - iq\Omega_p \rho_{da} \quad (2.148)$$

$$\dot{\rho}_{cb} = -i\left(\Delta_p - \frac{\omega_S}{2} + \Gamma_{cb}\right)\rho_{cb} + i\Omega_p \rho_{bb} + i\Omega_{IR}\rho_{db} \\ - iq\Omega_p \rho_{ca} - i\Omega_p \rho_{cc} \quad (2.149)$$

$$\dot{\rho}_{ac} = -(i\omega_S + \Gamma_{ac})\rho_{ac} - i\Omega_p \rho_{ab} + iq\Omega_p \rho_{bc} - i\Omega_{IR}\rho_{ad} + ik\Omega_{IR}\rho_{dc} \quad (2.150)$$

$$\dot{\rho}_{cd} = -i\left(\Delta_{IR} - \frac{\omega_S}{2} + \Gamma_{cd}\right)\rho_{cd} + i\Omega_p \rho_{bd} + i\Omega_{IR}\rho_{dd} \\ - i\Omega_{IR}\rho_{cc} - ik\Omega_{IR}\rho_{ca} \quad (2.151)$$

$$\dot{\rho}_{ad} = -i\left(\Delta_{IR} + \frac{\omega_S}{2} + \Gamma_{ad}\right)\rho_{ad} + iq\Omega_p \rho_{bd} + ik\Omega_{IR}\rho_{dd} \\ - ik\Omega_{IR}\rho_{aa} - i\Omega_{IR}\rho_{ac} \quad (2.152)$$

where $\omega_s = v_{ab} - v_{cb}$, $v_0 = \frac{v_{ab}+v_{cb}}{2}$, $v'_0 = \frac{v_{ad}+v_{cd}}{2}$ and $\Delta_p = v_0 - v_p$, $\Delta_{IR} = v'_0 - v_{IR}$. The population and dephasing decay rates are added phenomenologically in the above density matrix equations. The population decay rate for subband $|j\rangle$ (due to LO-phonon emission events) is denoted by γ_j. The total decay rates are given by:

$$\Gamma_{cb} = \gamma_c + \gamma_{cb}^{dph},$$
$$\Gamma_{ab} = \gamma_a + \gamma_{ab}^{dph},$$
$$\Gamma_{db} = \gamma_d + \gamma_{db}^{dph} \quad (\gamma_d = \gamma_{cd} + \gamma_{ad}), \tag{2.153}$$
$$\Gamma_{ac} = \gamma_a + \gamma_c + \gamma_{ac}^{dph}, \Gamma_{cd} = \gamma_c + \gamma_d + \gamma_{cd}^{dph} \quad \text{and}$$
$$\Gamma_{ad} = \gamma_a + \gamma_d + \gamma_{ad}^{dph}$$

In these expressions γ_{ij} (determined by electron–electron, interfaces roughness, and phonon scattering process) is the dephasing decay rate of the $|i\rangle \leftrightarrow |j\rangle$ transition. In this case the probe field interacts with both the sub band transitions $|b\rangle \leftrightarrow |c\rangle$ and $|b\rangle \leftrightarrow |a\rangle$ simultaneously with the Rabi frequencies Ω_p and $q\Omega_p$ ($q = \mu_{ab}/\mu_{cb}$ is dipole moment of relevant transition). On the other hand the terahertz-infrared field interacts with both $|d\rangle \leftrightarrow |c\rangle$ and $|d\rangle \leftrightarrow |a\rangle$ with the Rabi frequencies Ω_{IR} and $K\Omega_p$ (where $K = \mu_{ad}/\mu_{cd}$ is dipole moment of relevant transition). The optical susceptibility can be written as:

$$\chi = \frac{P}{\varepsilon_0 E} = \frac{2N_a \left(\wp_{ab}^2 \rho_{ab} + q \wp_{cb}^2 \rho_{cb} \right)}{\varepsilon_0 \Omega_p \hbar} \tag{2.154}$$

where Ω_p and N_a are the Rabi frequency of the probe field and the carrier density, respectively. The environment average thermal energy is considered as follows:

$$E_{thermal} = \frac{3}{2} KT \tag{2.155}$$

where K and T are the thermal energy, the Boltzmann constant and temperature in Kelvin, respectively. The effect of the above energy on the system operation is considered by adding this energy to the signal energy as follow:

$$E_T = E_{IR} + E_{thermal} \tag{2.156}$$

This section shows simulated results including transmission coefficient in proposed 4-subband quantum well structure. The transmission coefficient of probe field is proportional to the imaginary part of susceptibility [140]. Figure 2.70 shows the transmission coefficient of this structure. According to the matrix density equations, when the IR radiation is off, we will see two sharp absorption (Autler–Townes doublet) in the system transmission coefficient which are related to the transition, $|b\rangle \leftrightarrow |a\rangle$ and $|b\rangle \leftrightarrow |c\rangle$ respectively. The width of this Autler–Townes doublet is proportional to the splitting energy ($\langle 3|V_0|4\rangle$).

Figure 2.71 shows the effect of IR radiation on transmission coefficient of probe field in proposed asymmetric quantum well system. In the presence of IR radiation, a new sharp absorption is created in between the Autler–Townes doublet (for $K = 1$, $q = 1$ where K and q are the dipole moments). Solving the Schrödinger equation for this 4-subbands quantum well shows that the splitting energy is

2.6 Terahertz and Infrared Photodetector

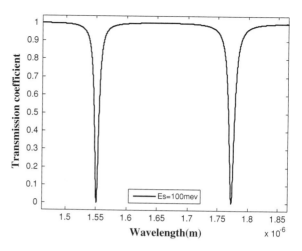

Fig. 2.70 Transmission coefficient versus wavelength for double asymmetric quantum wells systems ($\Omega_{IR} = 0\,s^{-1}, \Omega_p = 10^8\,s^{-1}, \Delta\Omega_{IR} = 0, E_S = 100\,\text{mev}, N_a = 10^{18}\,\text{cm}^{-3}, \Gamma_{ab} = \gamma_a + \gamma_{ab}^{dph} \approx 4.9\,\text{mev}, \Gamma_{cb} = \gamma_b + \gamma_{cb}^{dph} \approx 4.2\,\text{mev}, \Gamma_{ac} = \gamma_a + \gamma_c + \gamma_{ac}^{dph} \approx 8.3\,\text{mev}, K = \frac{\mu_{ad}}{\mu_{cd}} = 1, q = \frac{\mu_{ab}}{\mu_{cb}} = 1, T = 0\,K$)

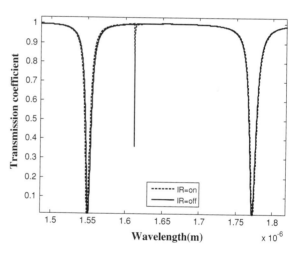

Fig. 2.71 Transmission coefficient versus wavelength for IR signal off (*solid line*) and IR signal on (*dash line*) with $K = 1, q = 1$ ($\Omega_{IR} = 20 * 10^{11}\,s^{-1}, \Omega_p = 10^8\,s^{-1}, \Delta\Omega_{IR} = 20\%, E_S = 100\,\text{mev}, N_a = 10^{18}\,\text{cm}^{-3}, \Gamma_{ab} = \gamma_a + \gamma_{ab}^{dph} \approx 4.9\,\text{mev}, \Gamma_{cb} = \gamma_b + \gamma_{cb}^{dph} \approx 4.2\,\text{mev}, \Gamma_{ac} = \gamma_a + \gamma_c + \gamma_{ac}^{dph} \approx 8.3\,\text{mev}, K = \frac{\mu_{ad}}{\mu_{cd}} = 1, q = \frac{\mu_{ab}}{\mu_{cb}} = 1, T = 0\,K$)

not so much (10–100 meV), so the magnitude of dipole moments K and q are closer to one.

Figure 2.72 shows the effect of IR intensity on the transmission coefficient of optical probe field. It is clear that the magnitude of this new absorption is sensitive to the infrared intensity. Therefore this asymmetric double quantum well structure can be used as a novel basic cell for terahertz-photodetectors.

Figure 2.73 shows the effect of IR-detuning on the transmission coefficient. When detuning is zero (IR field and probe field are coupled to the center of states $|a\rangle$ and $|b\rangle$) there is no absorption in the transmission coefficient. In fact due to symmetric case of system, the interaction of optical electromagnetically field with two states are the same and cancel out each other. Otherwise when the IR field and probe field are detuned, the new sharp absorption peaks are observed and their magnitudes are sensitive to the IR intensity.

Fig. 2.72 Transmission coefficient versus wavelength for different IR intensity with $K = 1, q = 1$ ($\Omega_p = 10^8 \text{ s}^{-1}, \Delta\Omega_{IR} = 20\%, E_S = 100 \text{ mev}, N_a = 10^{18} \text{ cm}^{-3}, \Gamma_{ab} = \gamma_a + \gamma_{ab}^{dph} \approx 4.9 \text{ mev}, \Gamma_{cb} = \gamma_b + \gamma_{cb}^{dph} \approx 4.2 \text{ mev}, \Gamma_{ac} = \gamma_a + \gamma_c + \gamma_{ac}^{dph} \approx 8.3 \text{ mev}, K = \frac{\mu_{ad}}{\mu_{cd}} = 1, q = \frac{\mu_{ab}}{\mu_{cb}} = 1, T = 0 \text{ K}$)

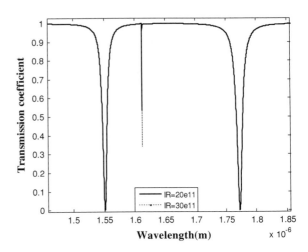

Fig. 2.73 Transmission coefficient versus wavelength for different IR detuning with $K = 1, q = 1$ ($\Omega_{IR} = 20 * 10^{11} \text{ s}^{-1}, \Omega_p = 10^8 \text{ s}^{-1}, E_S = 100 \text{ mev}, N_a = 10^{18} \text{ cm}^{-3}, \Gamma_{ab} = \gamma_a + \gamma_{ab}^{dph} \approx 4.9 \text{ mev}, \Gamma_{cb} = \gamma_b + \gamma_{cb}^{dph} \approx 4.2 \text{ mev}, \Gamma_{ac} = \gamma_a + \gamma_c + \gamma_{ac}^{dph} \approx 8.3 \text{ mev}, K = \frac{\mu_{ad}}{\mu_{cd}} = 1, q = \frac{\mu_{ab}}{\mu_{cb}} = 1, T = 0 \text{ K}$)

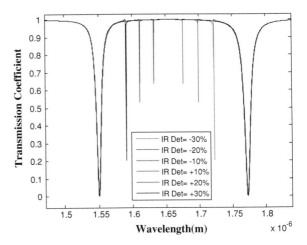

If we are able to change the dipole moment ratios, we can see some interesting cases. Figure 2.74 shows the effect of IR signal on the transmission coefficient for $K = 2.5$ and $q = 1$. It is clear that the new peak appears in center of spectrum and its magnitude is related to the IR intensity. In this case the interaction of electromagnetically field with two states is different and can not cancel out each other. Therefore a new absorption peak can be occurred.

There is another interesting case. Figure 2.75 shows the effect of IR radiation on the transmission coefficient when the IR field is resonant with the transition energy $|d\rangle \leftrightarrow |a\rangle$ or $|d\rangle \leftrightarrow |c\rangle$ (detuning = ±50%). We see a new sharp transparency in the spectrum where there was absorption. Figure 2.76 shows the magnitude of this new transparency is changed when the IR intensity changes.

Finally we investigate the effect of environment temperature on operation of the proposed quantum well structure. The environment thermal energy may be added

2.6 Terahertz and Infrared Photodetector

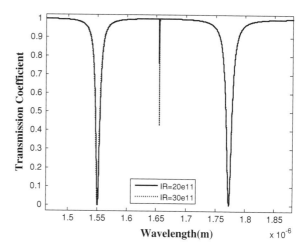

Fig. 2.74 Transmission coefficient versus wavelength with different IR intensity for $K = 2.5$, $q = 1$ ($\Omega_{IR} = 20 * 10^{11}\,s^{-1}$, $\Omega_p = 10^8\,s^{-1}$, $\Delta\Omega_{IR} = 0\%$, $E_S = 100\,mev$, $N_a = 10^{18}\,cm^{-3}$, $\Gamma_{ab} = \gamma_a + \gamma_{ab}^{dph} \approx 4.9\,mev$, $\Gamma_{cb} = \gamma_b + \gamma_{cb}^{dph} \approx 4.2\,mev$, $\Gamma_{ac} = \gamma_a + \gamma_c + \gamma_{ac}^{dph} \approx 8.3\,mev$, $K = \frac{\mu_{ad}}{\mu_{cd}} = 2.5$, $q = \frac{\mu_{ab}}{\mu_{cb}} = 1$, $T = 0\,K$)

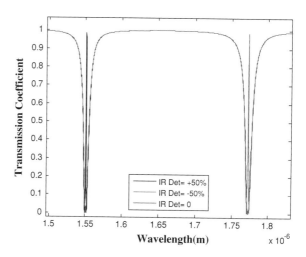

Fig. 2.75 Transmission coefficient versus wavelength for IR transition $|d\rangle \leftrightarrow |a\rangle$ and $|d\rangle \leftrightarrow |c\rangle$ (detuning = ±50%) ($\Omega_{IR} = 20 * 10^{11}\,s^{-1}$, $\Omega_p = 10^8\,s^{-1}$, $\Delta\Omega_{IR} = 50\%$, $E_S = 100\,mev$, $N_a = 10^{18}\,cm^{-3}$, $\Gamma_{ab} = \gamma_a + \gamma_{ab}^{dph} \approx 4.9\,mev$, $\Gamma_{cb} = \gamma_b + \gamma_{cb}^{dph} \approx 4.2\,mev$, $\Gamma_{ac} = \gamma_a + \gamma_c + \gamma_{ac}^{dph} \approx 8.3\,mev$, $K = \frac{\mu_{ad}}{\mu_{cd}} = 1$, $q = \frac{\mu_{ab}}{\mu_{cb}} = 1$, $T = 0\,K$)

to the incoming target IR signal and this effect inhibits the correct IR detection in the conventional photodetectors. However we show that this problem is not critical in our proposed EIT-based 4-subbands asymmetric quantum well structure. In the EIT-based structure the environment thermal energy cannot directly excite electrons (state $|b\rangle$ is populated but state $|d\rangle$ is not) but it causes some detuning in IR field. Figure 2.77 shows the effect of environment temperature on the transmission coefficient. If we couple IR without detuning, the environment temperature is added to IR signal and case a detuning (for $T = 0\,K \rightarrow 300\,K$, $\Delta\lambda \approx 80\,nm$). Now the effect of room temperature variations is considered. The 4 nm shift of wavelength for $20\,°C$ of room temperature variation is observed. It should be mentioned that when the refraction index of material increased ($n_{\text{refractive}} = 1 \rightarrow$

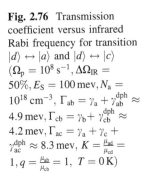

Fig. 2.76 Transmission coefficient versus infrared Rabi frequency for transition $|d\rangle \leftrightarrow |a\rangle$ and $|d\rangle \leftrightarrow |c\rangle$ ($\Omega_p = 10^8$ s^{-1}, $\Delta\Omega_{IR} = 50\%$, $E_S = 100$ mev, $N_a = 10^{18}$ cm^{-3}, $\Gamma_{ab} = \gamma_a + \gamma_{ab}^{dph} \approx 4.9$ mev, $\Gamma_{cb} = \gamma_b + \gamma_{cb}^{dph} \approx 4.2$ mev, $\Gamma_{ac} = \gamma_a + \gamma_c + \gamma_{ac}^{dph} \approx 8.3$ mev, $K = \frac{\mu_{ad}}{\mu_{cd}} = 1$, $q = \frac{\mu_{ab}}{\mu_{cb}} = 1$, $T = 0$ K)

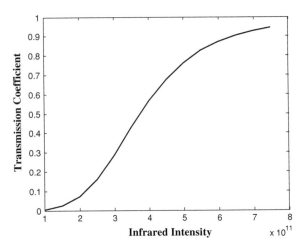

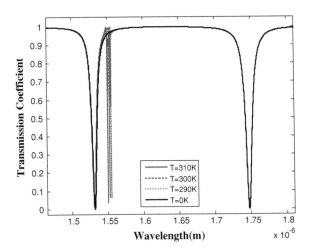

Fig. 2.77 Transmission coefficient versus wavelength for different room temperature with $K = 1$, $q = 1$ ($\Omega_{IR} = 20 * 10^{11}$ s^{-1}, $\Omega_p = 10^8$ s^{-1}, $\Delta\Omega_{IR} = 0\%$, $E_S = 100$ mev, $N_a = 10^{18}$ cm^{-3}, $\Gamma_{ab} = \gamma_a + \gamma_{ab}^{dph} \approx 4.9$ mev, $\Gamma_{cb} = \gamma_b + \gamma_{cb}^{dph} \approx 4.2$ mev, $\Gamma_{ac} = \gamma_a + \gamma_c + \gamma_{ac}^{dph} \approx 8.3$ mev, $K = \frac{\mu_{ad}}{\mu_{cd}} = 1$, $q = \frac{\mu_{ab}}{\mu_{cb}} = 1$, $T = 300$ K)

3.5 (for GaAs)), the wavelength temperature coefficient can be decreased ($\Delta\lambda/\Delta T = 0.2$ nm/°C $\rightarrow$ 60 pm/°C) that is so excellent from practical implementation point of view.

2.6.3 Terahertz Quantum Cascade Photodetector Based on Electromagnetically Induced Transparency

In spite of the numerous reports on QCDs especially in mid-infrared region, there is still lack of studies in terahertz quantum cascade devices since the thermal energy directly excites the ground state electrons to higher energy levels and in the

2.6 Terahertz and Infrared Photodetector

presence of transport ladder, leads to thermionic dark current [1, 27, 86, 135]. This problem deteriorates the performance of QCD.

A great number of the reported works on QCDs up to now are generally realized in III–V material system [3, 148]. However due to special features of GaN-based material system such as large conduction band offset in III-nitride heterostructures (which provides operation at much shorter wavelengths than other III–V QCDs), III-nitride heterostructures are highly suitable for implementation of such devices. On the other hand, the large lattice mismatch between AlN and GaN (as a deficiency) is not an impediment in QCDs since the barriers throughout the structure must be thin to obtain strong coupling between the wells.

Also, the improvement in the device performance due to the use of a phonon ladder is expected to be largely enhanced in the III-nitride material system, where the LO–phonon interaction is an order of magnitude stronger than in other III–V materials.

One can utilize the interference effect between the terahertz electromagnetic field and the short-wavelength probe field in a QCD active region to modify the absorption characteristic of probe field and achieve to coherence based photo detection [150]. Therefore, the incoming terahertz signal and the thermal energy do not directly excite ground state electrons and the thermionic dark current cancels out. In fact, we interpret the absorption of incoming terahertz signal as the absorption of short-wavelength probe signal by means of EIT Phenomenon [140, 150].

The EIT-based QCD considered in this section is a GaN/AlGaN heterostructure where the first quantum well of the period is n doped in order to populate the first energy level $E1$ in the conduction band with electrons (the nominal doping concentration of 5×10^{11} cm^{-2} has been considered). Figure 2.78 presents wave functions associated with each energy subband, in one period of the device. The layer widths in one period from left to right (active region and ladder) are: 20, 16, 10, 22, 10, 30, 20, 6, 20, 6, 20, 7, 20, 8, 20, 10, 20, 11, 20, 13, 20 and 15 Å, respectively.

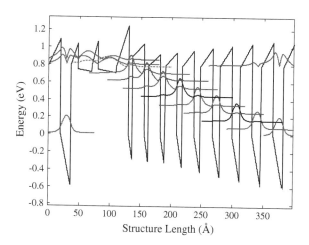

Fig. 2.78 Schematic of the conduction band profile, wave functions and associated energy levels of the EIT-based terahertz quantum cascade photodetector

Fig. 2.79 Simplified band diagram of the active region with schematically shown probe and terahertz signal wavelengths

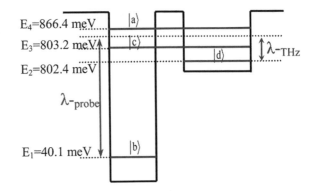

In Fig. 2.79 the active region of the QCD structure is schematically presented of which consist two coupled asymmetric quantum wells. The probe and THz signal wavelengths have been considered to be about 1.55 and 40 µm, respectively. It should be noted that the mid barrier inside the shallow well in Fig. 2.78 is placed to control the THz signal wavelength as a freedom degree of design.

In the interaction picture, the total Hamiltonian for this 4-subband quantum well structure can be written as [140]:

$$H = \hbar \begin{bmatrix} v_b & 0 & -\Omega_p e^{iv_p t} & -q\Omega_p e^{iv_p t} \\ 0 & v_d & -\Omega_{IR} e^{iv_{IR} t} & -k\Omega_{IR} e^{iv_{IR} t} \\ -\Omega_p e^{-iv_p t} & -\Omega_{IR} e^{-iv_{IR} t} & v_c & 0 \\ -q\Omega_p e^{-iv_p t} & -k\Omega_{IR} e^{-iv_{IR} t} & 0 & v_4 \end{bmatrix} \quad (2.157)$$

It is assumed that the Rabi frequencies of the probe signal (Ω_P) and THz-signal (Ω_{THz}) to be real. Using the density matrix formalism, one may describe the dynamic response of the proposed 4-subband quantum well structure. Under the electro-dipole and rotating-wave approximations (this method has described the results in experimental reports [142–144] and has been used in several theoretical papers [145–147]), the time-dependent density-matrix equations of motion can be obtained as [140]:

$$\dot{\rho}_{ab} = -[i(\Delta_p + \omega_S/2) + \Gamma_{ab}]\rho_{ab} + iq\Omega_p \rho_{bb} - iq\Omega_p \rho_{aa} \\ + ik\Omega_{IR}\rho_{db} - i\Omega_p \rho_{ac} \quad (2.158)$$

$$\dot{\rho}_{cb} = -[i(\Delta_p - \omega_S/2) + \Gamma_{cb}]\rho_{cb} + i\Omega_p \rho_{bb} + i\Omega_{IR}\rho_{db} \\ - iq\Omega_p \rho_{ca} - i\Omega_p \rho_{cc} \quad (2.159)$$

$$\dot{\rho}_{dc} = -[i(\Delta_{IR} + \omega_S/2) + \Gamma_{dc}]\rho_{dc} - i\Omega_p \rho_{db} - i\Omega_{IR}\rho_{dd} \\ + i\Omega_{IR}\rho_{cc} + ik\Omega_{IR}\rho_{ac} \quad (2.160)$$

2.6 Terahertz and Infrared Photodetector

$$\dot{\rho}_{db} = -i(\Delta_p + \Delta_{IR} + \Gamma_{db})\rho_{db} + i\Omega_{IR}\rho_{cb} + ik\Omega_{IR}\rho_{ab} \\ - i\Omega_p\rho_{dc} - iq\Omega_p\rho_{da} \tag{2.161}$$

$$\dot{\rho}_{da} = -[i(\Delta_{IR} - \omega_S/2) + \Gamma_{da}]\rho_{da} - iq\Omega_p\rho_{db} - ik\Omega_{IR}\rho_{dd} \\ + ik\Omega_{IR}\rho_{aa} + i\Omega_{IR}\rho_{ca} \tag{2.162}$$

$$\dot{\rho}_{ac} = -[i\omega_s + \Gamma_{ac}]\rho_{ac} - i\Omega_p\rho_{ab} + iq\Omega_p\rho_{bc} - i\Omega_{IR}\rho_{ad} + ik\Omega_{IR}\rho_{dc} \tag{2.163}$$

where $\omega_s = v_{ab} - v_{cb}$, $v_0 = (v_{ab} + v_{cb})/2$, $v'_0 = (v_{ad} + v_{cd})/2$, $\Delta_p = v_0 - v_p$, $\Delta_{THz} = v'_0 - v_{THz}$ and $q = \mu_{ab}/\mu_{cb}$, $K = \mu_{ad}/\mu_{cd}$. In the above density matrix equations the total decay rates are denoted by Γ_{ij}. In terms of the density matrix notation, the optical polarization is expressed as [140]:

$$P(z,t) = \left[\wp_{ab}(\rho_{ab(z,t)} + \text{c.c}) + \wp_{cb}(\rho_{cb(z,t)} + \text{c.c})\right] \tag{2.164}$$

where $\wp_{ij}$ is dipole moment matrix element between sates $|i\rangle$, $|j\rangle$. The linear susceptibility can be calculated from the optical polarization as follows [140]:

$$\chi_p^{(1)} = \frac{2N/\varepsilon_0\hbar\left[\wp_{ab}^2(L_1L_2 + \Omega_{IR}^2) + \wp_{cb}^2(-K\Omega_{IR}\Omega_{IR} + qL_3L_2 + qK\Omega_{IR}^2)\right]}{L_1L_2L_3 + L_3\Omega_{IR}^2 + L_1 + K^2\Omega_{IR}^2} \tag{2.165}$$

where $L_1 = [i(\Delta_p - \omega_s/2) + \Gamma_{cb}]$, $L_2 = [i(\Delta_p + \Delta_{THz}) + \Gamma_{db}]$, $L_3 = [i(\Delta_p + \omega_s/2) + \Gamma_{ab}]$, $K = \mu_{da}/\mu_{dc}$ and $q = \mu_{ab}/\mu_{cb}$.

The transmission coefficient of probe field is proportional to the imaginary part of susceptibility [149, 150].

Figure 2.80 shows the simulated transmission coefficient of probe field. According to the matrix density equations, when the terahertz radiation is off (Fig. 2.80a), two sharp absorption (like Autler–Townes doublet in atomic system) are appeared which are related to $|b\rangle \leftrightarrow |a\rangle$ and $|b\rangle \leftrightarrow |c\rangle$ transitions, respectively. The energy gap between two absorption peaks is proportional to the splitting energy (E_4–E_3). Figure 2.80b–d show the effect of terahertz radiation intensity on transmission coefficient. In the presence of terahertz radiation, a new sharp absorption is created between the two absorption peaks (interferential state). It is clear that the magnitude of this new absorption is sensitive to the terahertz intensity. Therefore this asymmetric double quantum well structure can be used as a novel active-region cell for terahertz photo-detection. In this feature, the THz signal does not manipulate the active region carriers directly (unlike the conventional THz quantum well photo-detectors) but the interference of the probe and THz signals modifies the absorption coefficient of the active region. This phenomenon may be described as the translation of the THz to probe short wavelength. Thus the cascade ladder coupled to the interferential state of the active region extracts the probe-excited carriers and transport them to consequent period.

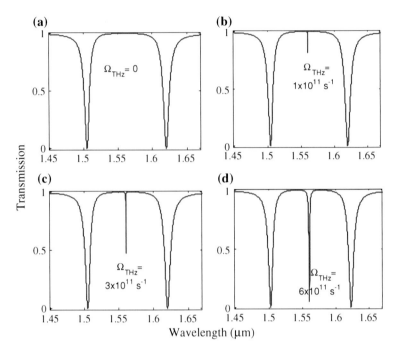

Fig. 2.80 Transmission coefficient of the active region versus wavelength for different THz-Rabi frequencies (Ω_{THz})

Fig. 2.81 Resistivity as a function of $1000/T$. The *inset* is a schematic of the three dominant transitions (highest transition rates in one period of the structure)

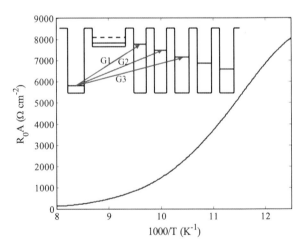

As previous discussions, in a photovoltaic detector, the zero voltage resistance, R_0A (which characterize the dark current-measured in the absence of illumination- and the deduced Johnson noise limited detectivity) should be calculated from $R_0A = k_B T/q^2 \sum_{i \in A} \sum_{j \in B} G_{ij}$, where the term G_{ij} is the phonon mediated global

Table 2.7 Quantities of the three largest transition rates for three sample temperatures in $m^{-2}s^{-1}$

Temperature (K)	G1	G2	G3
80	0.83×10^{17}	1.1×10^{13}	2.3×10^{8}
100	0.6×10^{18}	1.7×10^{14}	3.2×10^{9}
120	5.7×10^{18}	1.1×10^{15}	1.8×10^{10}

transition rate from a subband i of a cascade A to a subband j of a cascade B (in $m^{-2} s^{-1}$) [4, 5]. R_0A can usually be described with an activation energy E_a, corresponding to the energy of the transition responsible for electron transfer from one contact to the other.

Figure 2.81 presents the resistivity at R_0A as a function of $1,000/T$ where T is the temperature of the sample for the structure. The inset of the Fig. 2.68 schematically illustrates the three largest G_{ij} quantities ($G1$, $G2$ and $G3$) obtained in the simulation where the values are reported in Table 2.7 in 80, 100 and 120 K as sample temperatures.

References

1. Levine, B.F.: Quantum-well infrared photodetectors. J. Appl. Phys. **74**, R1–R81 (1993)
2. Graf, M.: Design and characterization of far- and mid-infrared quantum cascade detector. Trafford Publishing, Bloomington (2007)
3. Koeniguer, C., Dubois, G., Gomez, A., Berger, V.: Electronic transport in quantum cascade structures at equilibrium. Phys. Rev. B. **74**(23), 235325-1–235325-6 (2006)
4. Radovanović, J., Milanović, V., Ikonić, Z., Indjin, D., Harrison, P.: Electron–phonon relaxation rate. J. Appl. Phys. **97**, 103109-1–103109-5 (2005)
5. Ferreira, R., Bastard, G.: Evaluation of some scattering times for electrons in unbiased and biased signal- and multiple quantum well structures. Phys. Rev. B. **40**, 1074–1086 (1989)
6. Ozturk, E., Sokmen, I.: Intersubband transitions in quantum wells. Appl. Phys. D. **36**, 2457–2464 (2003)
7. Rosencher, E., Vinter, B.: Optoelectronics. Cambridge University Press, Cambridge (2002)
8. Beck, W.A.: Photoconductive gain and generation-recombination noise in multiple-quantum-well infrared detectors. Appl. Phys. Lett. **63**, 3589–3591 (1993)
9. Levine, B.F., Bethea, C.G., Choi, K.K., Walker, J., Malik, R.J.: Tunneling lifetime broadening of the quantum well intersubband photoconductivity spectrum. Appl. Phys. Lett. **53**, 231–233 (1988)
10. Suzuki, N., Iizuka, N., Kaneko, K.: Simulation of ultrafast GaN/AlN intersubband optical switches. IEICE Trans. Electron. **E88–C**, 342–348 (2005)
11. Liu, H.C.: Quantum dot infrared photodetector. Optoelectron. Rev. **11**, 1–5 (2003)
12. Su, X., Chakrabarti, S., Bhattacharya, P., Ariyawansa, G., Unil Perera, A.G.: A resonant tunneling quantum-dot infrared photodetector. IEEE J. Quantum Electron. **41**, 974–979 (2005)
13. Choi, K.K.: The physics of quantum well infrared photodetectors. World Scientific, Singapore (1997)
14. Gunapala, S.D., Bandara, S.V.: Quantum well infrared photodetector (QWIP) focal plane arrays. Semicond. Semimet. Ser. **62**, 1–83 (1999)
15. Steiner, T. (ed.): Semiconductor Nanostructures for Optoelectronic Applications. Artech House, Inc., Boston (2004)

16. Koeniguer, C., Gendron, L., Berger, V.: Analysis of performances of quantum cascade detectors. In: Proceedings of SPIE, vol. 5957, pp. 595704-1–595704-9 (2005)
17. Sarusi, G., Carbone, A., Gunapala, S.D., Liu, H.C. (eds.): High resistance narrow band quantum cascade photodetectors. In: Proceedings of the workshop on quantum well infrared photodetectors QWIP, Castello di Pavone, Torino, Italy, pp. 13–17 (2002)
18. Bastard, G.: Resonant carrier capture by semiconductor QW. Phys. Rev. B **33**, 1420–1423 (1986)
19. Brum, J.A., Weil, T., Nagle, J., Vinter, B.: Calculation of capture time of QW in graded-index separate-confinement heterostructures. Phys. Rev. B **34**, 2381–2384 (1986)
20. Rosencher, E., Vinter, B., Luc, F., Thibaudeau, L., Bois, P., Nagle, J.: Emission and capture of electrons in multiquantum-well structures. IEEE J. Quantum Electron. **30**, 2875–2888 (1994)
21. Liu, H.C.: Photoconductive gain mechanism of quantum-well intersubband infrared detectors. Appl. Phys. Lett. **60**, 1507–1509 (1992)
22. Kastalsky, A.: Photovoltaic in GaAs/AlGaAs. Appl. Phys. Lett. **52**, 1320–1322 (1988)
23. Goossen, K.W., Lyon, S.A.: Performance aspects of a QW detectors. J. Appl. Phys. **63**, 5149–5153 (1988)
24. Schneider, H., Larkins, E.C.: Space-charge effects in photovoltaic double barrier quantum well infrared detectors. Appl. Phys. Lett. **63**, 782–784 (1993)
25. Schonbein, C.: A GaAs/AlGaAs intersubband photodetectore operating at zero bias voltage. Appl. Phys. Lett. **68**, 973–975 (1996)
26. Schneider, H., Koidl, P., Schönbein, C., Ehret, S., Larkins, E.C., Bihlmann, G.: Capture dynamics and far-infrared response in photovoltaic quantum well intersubband photodetectors. Superlatt. Microstruct. **19**, 347–356 (1996)
27. Schneider, H., Liu, H.C.: Quantum Well Infrard Photodetectors. Springer, Berlin (2007)
28. Schneider, H., Schönbein, C., Walther, M., Schwarz, K., Fleissner, J., Koidl, P.: Photovoltaic QWIP the four-zone. Appl. Phys. Lett. **71**, 246–248 (1997)
29. Schneider, H., Koidl, P., Walther, M., Fleissner, J., Rehm, R., Diwo, E., Schwarz, K., Weimann, G.: Ten years of QWIP development at Fraunhofer IAF. Infrared Phys. Tech. **42**, 283–289 (2001)
30. Berger, V.: French patent, Détecteurs à cascade quantique. National reference number 0109754 (2001)
31. Koehler, R., et al.: Terahertz semiconductor-heterostructure laser. Nature **417**, 156–159 (2002)
32. Rostami, A., Motmaen, A., Baghban, H., Rasooli Saghai, H.: Dual-color mid-infrared quantum cascade photodetector in a coupled quantum well structure. In: Proceedings of SPIE-OSA-IEEE Asia communications and photonics, vol. 7631, pp. 76310L–76310L-10 (2009)
33. Ariyawansa, G., Perera, A.G.U., Su, X.H., Chakrabarti, S., Bhattacharya, P.: Multi-color tunneling quantum dot infrared photodetectores operation at room temperatute. Infrared Phys. Tech. **50**, 156–161 (2007)
34. Chamberlin, D., Robrish, P., Trutna, W., Scalari, G., Giovannini, M., Ajili, L., Faist, J., Beere, H., Ritchie, D.: Dual-wavelength THz imaging with quantum cascade lasers. Proc. SPIE **5727**, 107–114 (2005)
35. Ariyawansa, G., Apalkov, V., Perera, A.G.U., Matsik, S.G., Huang, G., Bhattacharya, P.: Bias-selectable tricolor tunneling quantum dot infrared photodetector for atmospheric windows. Appl. Phys. Lett. **92**, 111104-1–111104-3 (2008)
36. Kock, A., Gornik, E., Abstreiter, G., Bohm, G., Walther, M., Weimann, G.: Double wavelength selective GaAs/AlGaAs infrared detector device. Appl. Phys. Lett. **60**, 2011–2013 (1992)
37. Gravé, I., Shakouri, A., Kruze, N., Yariv, A.: Voltage-controlled tunable GaAs/AlGaAs multistack quantum well infrared photodetector. Appl. Phys. Lett. **60**, 2362–2364 (1992)
38. Liu, H.C., Li, J., Thompson, J.R., Wasilewski, Z.R., Buchanan, M., Simmons, J.G.: Multicolor voltage-tunable quantum-well infrared photodetector. IEEE Electron Device Lett. **14**, 566–568 (1993)

39. Tidrow, M.Z., Chiang, J.C., Li, S.S., Bacher, K.: A high strain two-stack two-color quantum well infrared photodetector. Appl. Phys. Lett. **70**, 859–861 (1997)
40. Tsai, K.L., Chang, K.H., Lee, C.P., Huang, K.F., Tsang, J.S., Chen, H.R.: Two-color infrared photodetector using GaAs/AlGaAs and strained InGaAs/AlGaAs multiquantum wells. Appl. Phys. Lett. **62**, 3504–3506 (1993)
41. Wang, Y.H., Chiang, J.C., Li, S.S., Ho, P.: A GaAs/AlAs/AlGaAs and GaAs/AlGaAs stacked quantum well infrared photodetector for 3–5 and 8–14 μm detection. J. Appl. Phys. **76**, 2538–2540 (1994)
42. Zhang, Y., Jiang, D.S., Xia, J.B., Cui, L.Q., Song, C.Y., Zhou, Z.Q., Ge, W.K.: A voltage-controlled tunable two-color infrared photodetector using GaAs/AlAs/GaAlAs and GaAs/GaAlAs stacked multiquantum wells. Appl. Phys. Lett. **68**, 2114–2116 (1996)
43. Beck, W.A., Faska, T.S.: Current status of quantum well focal plane arrays. In: Proceedings of the SPIE, infrared technology and applications XXII, vol. 2744, pp. 193–206 (1996)
44. Berger, V., Vodjdani, N., Bois, P., Vinter, B., Delaitre, S.: Switchable bicolor infrared detector using an electron transfer infrared modulator. Appl. Phys. Lett. **61**, 1898–1900 (1992)
45. Martinet, E., Rosencher, E., Luc, F., Bois, P., Costard, E., Delaitre, S.: Switchable bicolor (5.5–9.0 μm) infrared detector using asymmetric GaAs/AlGaAs multiquantum well. Appl. Phys. Lett. **61**, 246–248 (1992)
46. Chiang, J.C., Li, S.S., Tidrow, M.Z., Ho, P., Tsai, M., Lee, C.P.: A voltage-tunable multicolor triple-coupled InGAAS/GaAs/AlGaAs quantum-well infrared photodetector for 8–12 μm detection. Appl. Phys. Lett. **69**, 2412–2414 (1996)
47. Schneider, H., Liu, H.C., Winnerl, S., Song, C.Y., Drachenko, O., Walther, M., Faist, J., Helm, M.: Quadratic detection with two-photon quantum well infrared photodetectors. Infrared Phys. Tech. **52**, 419–423 (2009)
48. Schneider, H., Liu, H.C., Winnerl, S., Song, C.Y., Walther, M., Helm, M.: Terahertz two-photon quantum well infrared photodetector. Opt. Express **17**(15), 12279–12284 (2009)
49. Rosencher, E., Fiore, A., Vinter, B., Berger, V., Bois, P.h., Nagle, J.: Quantum engineering of optical nonlinearities. Science **271**, 168–173 (1996)
50. Schneider, H., Maier, T., Liu, H.C., Walther, M., Koidl, P.: Ultra-sensitive femtosecond two-photon detector with resonantly enhanced nonlinear absorption. Opt. Lett. **30**, 287–289 (2005)
51. Maier, T., Schneider, H., Walther, M., Koidl, P., Liu, H.C.: Resonant two-photon photoemission in quantum well infrared photodetectors. Appl. Phys. Lett **84**, 5162–5164 (2004)
52. Schneider, H., Liu, H.C.: Quantum Well Infrared Photodetectors: Physics and Applications. Springer, Berlin (2006)
53. Cristea, P., Fedoryshyn, Y., Holzman, J.F., Robin, F., Jäckel, H., Müller, E., Faist, J.: Tuning the intersubband absorption in strained AlAsSb/InGaAs quantum wells towards the telecommunications wavelength range. J. Appl. Phys. **100**, 116104-1–116104-3 (2006)
54. Luo, H., Liu, H.C., Song, C.Y., Wasilewski, Z.R.: Background-limited terahertz quantum-well photodetector. Appl. Phys. Lett. **86**, 231103-1–231103-3 (2005)
55. Tonouchi, M.: Cutting-edge terahertz technology. Nat. Photon. **1**, 97–105 (2007)
56. Liu, H.C., Song, C.Y., Wasilewski, Z.R., Gupta, J.A., Buchanan, M.: Fano resonance mediated by intersubband-phonon coupling. Appl. Phys. Lett. **91**, 131121-1–131121-3 (2007)
57. Schneider, H., Drachenko, O., Winnerl, S., Helm, M., Walther, M.: Quadratic autocorrelation of free-electron laser radiation and photocurrent saturation in two-photon quantum-well infrared photodetectors. Appl. Phys. Lett. **89**, 133508-1–133508-3 (2006)
58. Stier, O., Grundmann, M., Bimberg, D.: Electronic and optical properties of strained quantum dots modeled by 8-band k.p theory. Phys. Rev. B **59**, 5688–5701 (1999)
59. Sun, S.J., Chang, Y.C.: Modeling self-assembled quantum dots by the effective bond-orbital method. Phys. Rev. B **62**, 13631–13640 (2000)

60. Ryzhii, V.: Negative differential photoconductivity in quantum-dot infrared photodetectors. Appl. Phys. Lett. **78**, 3346–3348 (2001)
61. Boucaud, P., Sauvage, S.: Infrared photodetection with semiconductor self-assembled quantum dots. C. R. Phys. **4**, 1133–1154 (2003)
62. Xu, S.J., et al.: Characteristics of InGaAs quantum dot infrared photodetectors. Appl. Phys. Lett. **73**, 3153–3155 (1998)
63. Blick, R.H., Haug, R.J., Weis, J., Pfannkuche, D., Klitzing, K.V., Eberl, K.: Single-electron tunneling through a double quantum dot: the artificial molecule. Phys. Rev. B **53**, 7899–7902 (1996)
64. Sikorski, C., Merkt, U.: Spectroscopy of electronic states in InSb quantum dots. Phys. Rev. Lett. **62**, 2164–2167 (1989)
65. Demel, T., Heitmann, D., Grambow, P., Ploog, K.: Nonlocal dynamic response and level crossings in quantum-dot structures. Phys. Rev. Lett. **64**, 788–791 (1990)
66. Stranski, I.N., Krastanow, L.: Zur Theorie der orientierten Ausscheidung von Ionenkristallen aufeinander. Sitzungsber. Akad. Wiss. Wien, Math. Naturwiss. Kl Abt. 2B **146**, 797–810 (1938)
67. Goldstein, L., Glas, F., Marzin, J.Y., Charasse, M.N., Le Roux, G.: Growth by molecular beam epitaxy and characterization of InAs/GaAs strained-layer superlattices. Appl. Phys. Lett. **47**, 1099–1101 (1985)
68. Mo, Y.W., Savage, D.E., Swartzentruber, B.S., Lagally, M.G.: Kinetic pathway in Stranski–Krastanov growth of Ge on Si(001). Phys. Rev. Lett. **65**, 1020–1023 (1990)
69. Gérard, J.M., Marzin, J.Y., Zimmermann, G., Ponchet, A., Cabrol, O., Barrier, D., Jusserand, B., Sermage, B.: InAs/GaAs quantum boxes obtained by self-organized growth: intrinsic electronic properties and applications. Solid State Electron. **40**, 807–814 (1996)
70. Grundmann, M., Stier, O., Bimberg, D.: InAs/GaAs pyramidal quantum dots: strain distribution, optical phonons, and electronic structure. Phys. Rev. B **52**, 11969–11981 (1995)
71. Lee, H., Lowe-Webb, R., Yang, W., Sercel, P.C.: Determination of the shape of self-organized InAs/GaAs quantum dots by reflection high energy electron diffraction. Appl. Phys. Lett. **72**, 812–814 (1998)
72. Haller, E.E.: Advanced far-infrared detectors. Infrared Phys. **35**, 127–146 (1994)
73. Suzuki, M., Tonouchi, M.: Fe-implanted InGaAs photoconductive terahertz detectors triggered by 1.56 μm femtosecond optical pulses. Appl. Phys. Lett. **86**, 163504-1–163504-3 (2005)
74. Rinzan, M.B.M., Perera, A.G.U., Matsik, S.G., Liu, H.C., Wasilewski, Z., Buchanan, M.: AlGaAs emitter/GaAs barrier terahertz detector with a 2.3 THz threshold. Appl. Phys. Lett. **86**, 071112-1–071112-3 (2005)
75. Lü, J., Shur, M.S.: Terahertz detection by high-electron-mobility transistor: enhancement by drain bias. Appl. Phys. Lett. **78**, 2587–2588 (2001)
76. Liu, H.C., Song, C.Y., Spring Thorpe, A.J., Cao, J.C.: Terahertz quantum-well photodetector. Appl. Phys. Lett. **84**, 4068–4070 (2004)
77. Phillips, J., Bhattacharya, P., Kennerly, S.W., Beekman, D.W., Dutta, M.: Self-assembled InAs–GaAs quantum-dot intersubband detectors. IEEE J. Quantum Electron. **35**, 936–943 (1999)
78. Kim, E., Madhukar, A., Ye, Z., Campbell, J.C.: High detectivity InAs quantum dot infrared photodetectors. Appl. Phys. Lett. **84**, 3277–3279 (2004)
79. Krishna, S., Raghavan, S., Von Winckel, G., Stintz, A., Ariyawansa, G., Matsik, S.G., Perera, A.G.U.: Three-color (λ_{p1} 3.8 μm, λ_{p2} 8.5 μm, and λ_{p3} 23.2 μm) InAs/InGaAs quantum-dots-in-a-well detector. Appl. Phys. Lett. **83**, 2745–2747 (2003)
80. Chakrabarti, S., Stiff-Roberts, A.D., Bhattacharya, P., Gunapala, S.D., Bandara, S., Rafol, S.B., Kennerly, S.W.: High-temperature operation of InAs/GaAs quantum dot infrared photodetectors with large responsivity and detectivity. IEEE Photonics Technol. Lett. **16**, 1361–1363 (2004)

References

81. Berryman, K.W., Lyon, S.A., Segev, M.: Mid-infrared photoconductivity in InAs quantum dots. Appl. Phys. Lett. **70**, 1861–1863 (1997)
82. Liu, J.L., Wu, W.G., Balandin, A., Jin, G.L., Wang, K.L.: Intersubband absorption in boron-doped multiple Ge quantum dots. Appl. Phys. Lett. **74**, 185–187 (1999)
83. Chua, Y.C., Decuir, E.A., Passmore, B.S., Sharif, K.H., Manasreha, M.O., Wang, Z.M., et al.: Tuning In0.3Ga0.7As/GaAs multiple quantum dots for long-wavelength infrared detectors. Appl. Phys. Lett. **85**, 1003–1005 (2004)
84. Chakrabarti, S., Bhattacharya, P., Stiff-Roberts, A.D., Lin, Y.Y., Singh, J., Lei, Y., et al.: Intersubband absorption in annealed InAs/GaAs quantum dots: a case for polarization-sensitive infrared detection. J. Phys. D Appl.Phys. **36**, 1794–1797 (2003)
85. Barseghyan, M.G., Kirakosyan, A.A.: Light absorption by a two-dimensional quantum dot superlattice. Physica E **27**, 474–480 (2005)
86. Etteh, N.E.I., Harrison, P.: Carrier scattering approach to the origins of dark current in mid and far-infrared (terahertz) quantum-well intersubband photodetectors (QWLPs). IEEE J. Quantum Electron. **37**, 672–675 (2001)
87. Phillips, J., Kamath, K., Bhattacharrya, P.: Characteristics of InAs/AlGaAs self-organized quantum dot modulation doped field effect transistors. Appl. Phys. Lett. **72**, 3509–3511 (1998)
88. Kim, S., Mohseni, H., Erdtmann, M., Michel, E., Jelen, C., Razeghi, M.: Growth and characterization of InGaAs/InGaP quantum dots for midinfrared photoconductive detector. Appl. Phys. Lett. **73**, 963–965 (1998)
89. Pan, D., Towe, E., Kennerly, S.: A five-period normal-incidence (In, Ga)As/GaAs quantum-dot infrared photodetector. Appl. Phys. Lett. **75**, 2719–2721 (1999)
90. Kim, J.W., Oh, J.E., Hong, S.C., Park, C.H., Yoo, T.K.: Room temperature far infrared ($8 \sim 10$ μm) photodetectors using self-assembled InAs quantum dots with high detectivity. IEEE Electron. Device Lett. **21**, 329–331 (2000)
91. Liu, H.C., Gao, M., McCaffrey, J., Wasilewski, Z.R., Fafard, S.: Quantum dot infrared photodetectors. Appl. Phys. Lett. **78**, 79–81 (2001)
92. Stiff, A.D., Krishna, S., Bhattacharya, P., Kennerly, S.W.: Normal-incidence, high-temperature, mid-infrared, InAs-GaAs vertical quantum-dot infrared photodetector. IEEE J. Quantum Electron. **37**, 1412–1419 (2001)
93. Stiff, A.D., Krishna, S., Bhattacharya, P., Kennerly, S.: High-detectivity, normal-incidence, mid-infrared ($\lambda \sim 4$ μm) InAs/GaAs quantum-dot detector operating at 150 K. Appl. Phys. Lett. **79**, 421–423 (2001)
94. Ye, Z., Campbell, J.C., Chen, Z., Kim, E., Madhukar, A.: Normal-incidence InAs self-assembled quantum-dot infrared photodetectors with a high detectivity. IEEE J. Quantum Electron. **38**, 1234–1237 (2002)
95. Ye, Z., Campbell, J.C., Chen, Z., Kim, E.T., Madhukar, A.: InAs quantum dot infrared photodetectors with $In_{0.15}Ga_{0.85}As$ strain–relief cap layears. J. Appl. Phys. **92**, 7462–7468 (2002)
96. Lee, S.W., Hirakawa, K., Shimada, Y.: Bound-to-continuum intersubband photoconductivity of self-assembled InAs quantum dots in modulation-doped heterostructures. Appl. Phys. Lett. **75**, 1428–1430 (1999)
97. Chu, L., Zrenner, A., Bichler, M., Abstreiter, G.: Quantum-dot infrared photodetector with lateral carrier transport. Appl. Phys. Lett. **79**, 2249–2251 (2001)
98. Hsu, B.C., Chang, S.T., Chen, T.C., Kuo, P.S., Chen, P.S., Pei, Z., et al.: A high efficient 820 nm MOS Ge quantum dot photodetector. IEEE Electron. Device Lett. **24**, 318–320 (2003)
99. Su, X.H., Yang, J., Bhattacharya, P., Ariyawansa, G., Perera, A.G.U.: Terahertz detection with tunneling quantum dot intersublevel photodetector. Appl. Phys. Lett. **89**, 031117-1–031117-3 (2006)
100. Esaev, D.G., Rinzan, M.B.M., Matsik, S.G., Perera, A.G.U., Liu, H.C., Zvonkov, B.N., Gavrilenko, V.I., Belyanin, A.A.: High performance single emitter homojunction interfacial workfunction far infrared detectors. J. Appl. Phys. **95**, 512–519 (2004)

101. Hofer, S., Hirner, H., Bratschitsch, R., Strasser, G., Unterrainer, K.: Photoconductive response of InAs/GaAs quantum dot stacks. Physica E **13**, 190–193 (2002)
102. Grundmann, M., Weber, A., Goede, K., Ustinov, V.M., Zhukov, A.E., Ledentsov, N.N., Kop'ev, P.S., Alferov, Z.I.: Midinfrared emission from near-infrared quantum-dot lasers. Appl. Phys. Lett. **77**, 4–6 (2000)
103. Wingreen, N.S., StaLord, C.A.: Quantum-dot cascade laser: proposal for an ultralow-threshold semiconductor laser. J. Quantum Electron. **33**, 1170–1173 (1997)
104. Bayer, M., Hawrylak, P., Hinzer, K., Fafard, S., Korkusinski, M., Wasilewski, R., Stern, O., Forchel, A.: Coupling and entangling of quantum states in quantum dot molecules. Science. **291**, 451–453 (2001)
105. Bastard, G.: Theoretical investigations of superlattice band structure in the envelope-function approximation. Phys. Rev. B **25**, 7584–7597 (1982)
106. Pan, D., Towe, E., Kennerly, S.: Normal-incidence intersubband (In, Ga)As/GaAs quantum dot infrared photodetectors. Appl. Phys. Lett. **73**, 1937–1939 (1998)
107. Chu, L., Zrenner, A., Bohm, G., Abstreiter, G.: Normal-incident intersubband photocurrent spectroscopy on InAs/GaAs quantum dots. Appl. Phys. Lett. **75**, 3599–3601 (1999)
108. Yu, L.S., Li, S.S.: A metal grating coupled bound-to-miniband transition GaAs multiquantum well/superlattice infrared detector. Appl. Phys. Lett. **59**, 1332–1334 (1991)
109. Yu, L.S., Wang, H., Li, S.S., Ho, P.: Low dark current step-bound-to-miniband transition InGaAs/GaAs/AlGaAs multiquantum-well infrared detector. Appl. Phys. Lett. **60**, 992–994 (1992)
110. Carr, G.L., Martin, M.C., Mckinney, W.R., Jordan, K., Neil, G.R., Williams, G.P.: High-power terahertz radiation from relativistic electrons. Nat. **420**, 153–156 (2002)
111. Sherwin, M.: Terahertz power. Nat. **420**, 131–133 (2002)
112. Mittleman, D.M., Jacobsen, R.H., Nuss, M.C.: T-Ray imaging'. IEEE J. Select. Topics. Quantum Eletronics. **2**, 679–692 (1996)
113. Siegel, P.H.: Terahertz technology in biology and medicine. IEEE Tran. Microwave. Theory and Techniques **52**, 2438–2447 (2004)
114. Siegel, P.H.: THz applications for outer and inner space. In: 17th International Zurich Symposium on Electromagnetic Compatibility, 1–4 (2006)
115. Dai, J.H., Lee, J.H., Lee, S.C.: Annealing effect on the formation of In(Ga)As quantum rings from InAs quantum dots. IEEE Photon. Technol. Lett. **20**, 165–167 (2008)
116. Lee, J.H., Dai, I.H., Chang, Y.T., Chan, C.F., Lee, S.C.: In(Ga)As quantum ring terahertz photodetector. Solid State Devices and Materials (SSDM 2008). Ibaraki, Japan (2008)
117. Kochman, B., Stiff-Roberts, A.D., Chakrabarti, S., Phillips, J.D., Krishna, S., Singh, J., et al.: Absorption, carrier lifetime, and gain in InAs–GaAs quantum-dot infrared photodetectors. IEEE J. Quantum Electron. **39**, 459–467 (2003)
118. Ryzhii, V., Khmyrova, I., Ryzhii, M., Mitin, V.: Comparison of dark current, responsivity and detectivity in different intersubband infrared photodetectors. Semicond. Sci. Technol. **19**, 8–16 (2004)
119. Rostami, A., Rasooli Saghai, H.: A novel proposal for ultra-high optical nonlinearity in GaN/AlGaN spherical centered defect quantum dot (SCDQD). Microelectron. J. **38**, 342–351 (2007)
120. Rostami, A., Rasooli Saghai, H., Sadoogi, N., Baghban, H.: Proposal for ultra-high performance infrared quantum dot. Opt. Express **16**, 2752–2763 (2008)
121. Rostami, A., Rasooli Saghai, H., Baghban, H.: A proposal for enhancement of absorption coefficient and electroabsorption properties in GaN/AlGaN centered defect quantum box (CDQB) nanocrystal. Phys. B **403**, 2789–2796 (2008)
122. Rostami, A., Rasooli Saghai, H., Baghban, H.: A proposal for enhancement of optical nonlinearity in GaN/AlGaN centered defect quantum box (CDQB) nanocrystal. Solid State Electron. **52**, 1075–1081 (2008)
123. Sadoogi, N., Rasooli Saghai, H., Rostami, A., Ghafoori Fard, H.: Electron transport in array of centered defect quantum dots. Phys. E **41**, 269–277 (2008)

124. Harrison, P., Gadir, M.A., Etteh, N.E.I., Soref, R.A.: The physics of THz QWIPs. In: Proceedings of the 9th international conference on THz electronics, the 10th IEEE international conference on terahertz electronics (2002)
125. Rasooli Saghai, H., Sadoogi, N., Rostami, A., Baghban, H.: Ultra-high detectivity room temperature THZ IR-photodetector based on resonant tunneling spherical centered defect quantum dot (RT-SCDQD). Optic. Comm. **282**, 3499–3508 (2009)
126. Rostami, A., Rasooli Saghai, H., Baghban Asghari Nejad, H., Sadoogi, N.: Tailoring of quantum dot basic cell towards high detectivity THz-IR photodetector. In: Luo, Y., Buus, J., Koyama, F., Lo, Y.-H. (eds.) Optoelectronic Materials and Devices III. Proc. of SPIE, vol. 7135, pp. 71352Z-1–71352Z-9 (2008)
127. Arfken, G.B., Weber, H.J.: Mathematical Methods for Physicists. Harcourt Academic Press, San Diego (2001)
128. Peyghambarian, N., Koch, S.W., Mysyrowicz, A.: Introduction to Semiconductor Optics. Prentice-Hall, Englewood Cliffs (1993)
129. Basu, P.K.: Theory of Optical Processes in Semiconductors. Clarendon Press, Oxford (1997)
130. Asgari, A., Kalafi, M., Faraone, L.: The effects of GaN capping layer thickness on two-dimensional electron mobility in GaN/AlGaN/GaN heterostructures. Phys. E **25**, 431–437 (2005)
131. Guo, K.X., Yu, Y.B.: Nonlinear optical susceptibilities in Si/SiO_2 parabolic quantum dots. Chin. J. Phys. **43**, 932–940 (2005)
132. Liu, J., Bai, J., Xiong, G.: Studies of the second-order nonlinear optical susceptibilities of GaN/AlGaN quantum well. Phys. E **23**, 70–74 (2004)
133. Adawi, A.M., Zibik, E.A., Wilson, L.R., Lemaitre, A., Cockburn, J.W., Skolnick, M.S., Hopkinson, M., et al.: Strong in-plane polarised intersublevel absorption in vertically aligned InGaAs/GaAs quantum dots. Appl. Phys. Lett. **82**, 3415–3417 (2003)
134. Kerr, W.E., Pancholi, A., Stoleru, V.G.: Quantum dot molecules: a potential pathway towards terahertz devices. Phys. E. **35**, 139–145 (2006)
135. Paiella, R.: Intersubband Transitions in Quantum Structures. McGraw-Hill, NY (2006)
136. Zyaei, M., Rasooli Saghai, H., Abbasian, K., Rostami, A.: Long wavelength infrared photodetector design based on electromagnetically induced transparency. Optic. Comm. **281**, 3739–3747 (2008)
137. Rostami, A., Zyaei, M., Rasooli Saghi, H.: Room temperature terahertz quantum well infrared photodetector based on electromagnetically induced transparency (EIT). In: Proceedings of the IASTED International Conference, Greece Nanotechnology and Application (NANA), vol. 615-050, pp. 95–100 (2008)
138. Rostami, A., Zyaei, M., Rasooli Saghai, H., Sharifi, F.J.: Terahertz asymmetric quantum well infrared photodetector design based on electromagnetically induced transparency. In: Optomechatronic technologies, SPIE, vol. 7266, pp. 72660Z-1–72660Z-9 (2008)
139. Harris, S.E.: Electromagnetically induced transparency. Phys. Today **50**(7), 36–42 (1997)
140. Scully, M.O., Zubairy, M.S.: Quantum Optics. Cambridge University Press, Cambridge (1997)
141. Phillips, C.C., Paspalakis, E., Serapiglia, G.B., Sirtori, C., Vodopyanov, K.L.: Observation of electromagnetically induced transparency and measurements of subband dynamics in a semiconductor quantum well. Phys. E **7**, 166–173 (2000)
142. Faist, J., Capasso, F., Sirtori, C., West, K., Pfeiffer, L.N.: Controlling the sign of quantum interference by tunnelling from quantum wells. Nature **390**, 589–591 (1997)
143. Schmidt, H., Campman, K.L., Gossard, A.C., Imamoglu, A.: Tunneling induced transparency: fano interference in intersubband transitions. Appl. Phys Lett. **70**, 3455–3457 (1997)
144. Frogley, M.D., Dynes, J.F., Beck, M., Faist, J., Phillips, C.C.: Gain without inversion in semiconductor nanostructures. Nat. Mater. **5**, 175–178 (2006)
145. Schmidt, H., Imamoglu, A.: Nonlinear optical devices based on a transparency in semiconductor intersubband transitions. Optic. Commun. **131**, 333–338 (1996)

146. Lee, C.R., Li, Y.C., Men, F.K., Pao, C.H., Tsai, Y.C., Wang, J.F.: Model for an inversionless two-color laser. Appl. Phys. Lett. **86**, 201112-1–201112-3 (2004)
147. Dyns, J.F., Frogley, M.D., Rodger, J., Phillips, C.C.: Optically mediated coherent population trapping in asymmetric semiconductor quantum wells. Phys. Rev. B **72**, 085323-1–085323-7 (2005)
148. Giorgetta, F.R., Baumann, A.E., Hofstetter, D., Manz, C., Yang, Q., Köhler, K., Graf, M.: InGaAs/AlAsSb quantum cascade detectors operating in the near infrared. Appl. Phys. Lett. **91**, 111115-1–111115-3 (2007)
149. Yelin, S.F., Hemmer, P.R.: Resonantly enhanced nonlinear optics in semiconductor quantum wells: an application to sensitive infrared detection. Phys. Rev. A **66**, 013803-1–013803-5 (2002)
150. Fleischhauer, M., Imamoglu, A., Marangos, J.P.: Electromagnetically induced trans

Chapter 3
Terahertz and Infrared Quantum Cascade Lasers

3.1 Introduction

The generation of mid-infrared and terahertz portion of the optical spectrum using the Quantum Cascade Laser (QCL) technology has the potential of making cheap, powerful optical, room temperature sources. In the mid-infrared spectral region, where continuous wave room temperature operation of the QCL devices was achieved, the main goal will be to further broaden the frequency range over which these high performances are achieved. Other important topic is the developing of devices with a very large active broadband region, with tuning range of more than 250 cm^{-1} for a laser emission centered at 1000 cm^{-1}. However, the overall level of performance of the THz QCL's (higher operating temperatures and longer wavelengths) in comparison to mid-infrared is much lower with the maximum known operating temperature and wavelength still being 160 K and 180 μm (1.7 THz) respectively. For this reason, the focus is to find solutions for optical laser cavity. Finally, as far as the photonic side is concerned, the concentration is on the realization of waveguides and resonators based on the sandwiching technique used for two metallic layers surface plasma.

Section 3.2 covers the QCL principles. Section 3.3 discuses about the terahertz QCLs. In the Sect. 3.4, the analysis of transport properties of THz QCLs will be introduced. Section 3.5 gives the brief review on high power THz-QCL. Dual-wavelength generation based on monolithic THz-IR QCL with containing two different structures will be introduced in Sect. 3.6.

3.2 Quantum Cascade Laser Principles

Impractical extension of bipolar laser technology to the far-infrared and terahertz region due to lack of materials with band-gap less than 40 meV motivated the development of new generation of lasers where the radiative transitions take place entirely within the conduction band between quantized states in heterostructure quantum wells. The photon energy that results from an intersubband transition can be chosen by engineering the thicknesses of the quantum wells and barriers, which make such structures ideal for the generation of long-wavelength radiation. In 1994, Faist et al. demonstrated the first intersubband laser, designed to emit at 4.3 μm and grown by molecular beam epitaxy and this invention was named QCL. One of the important characteristic of QCLs is the use of periodic multiple quantum-well (MQW) regions such that one carrier leads to generation of multiple photons as it is transported through the repeated regions. QCLs can cover a wide wavelength range of 3–215 μm (with the aid of magnetic field operation in higher frequencies). These lasers have so far primarily been realized in the conduction band of n-doped InGaAs/InAlAs, GaAs/AlGaAs and InGaAs/AlAsSb heterostructures and lasing around $\lambda \sim 10$ μm has been obtained in all of these three material systems which shows the versatility of the QCL concept. Compared to conventional interband lasers, QCLs have the following advantages: The emission wavelength is primarily a function of the QW thickness thus, they can be designed to emit at any wavelength over an extremely wide range using the same combination of materials in the active region. QCLs are designed based on a cascade of identical stages (typically 20–50), allowing one electron to emit many photons, emitting more optical power and finally, intersubband transitions are characterized by an ultrafast carrier dynamics and a small linewidth enhancement factor, with both features being expected to have significant impact on laser performance. In this section, before discussing the operation and development of QCLs, the optical properties of semiconductor heterostructures will briefly described.

3.2.1 Radiative and Non-radiative Transitions in Semiconductor Heterostructures

Since understanding the basic characteristics and optical properties of semiconductor lasers requires relatively deep insight to the carrier relaxation mechanism, it is worthy to study the two important types of electron transitions in semiconductors known as radiative and non-radiative transitions which are characterize by emitting a photon and phonons (especially longitudinal optical or LO-phonons are of more interest) respectively. These types of transitions are illustrated by corrugated and right arrows in Fig. 3.1 respectively.

3.2.1.1 Radiative Transitions

Transitions rate between an initial $|i\rangle$ and a final $|f\rangle$ states in quantum wells can defined by Fermi's golden rule as:

$$\wp_{if} = \frac{2\pi}{\hbar} |\langle \psi_f | W | \psi_i \rangle|^2 \delta(E_f - E_i \pm \hbar\omega), \tag{3.1}$$

3.2 Quantum Cascade Laser Principles

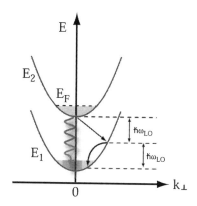

Fig. 3.1 First two subband energies in the in-plane wave vector $k_\perp$ plane. The *right arrow* corresponds to non-radiative transition assisted by the optical phonon emission, and the *corrugated arrow* corresponds to radiative transition assisted by photons emission [88]

where W is known as interaction Hamiltonian given by:

$$W(\vec{r},t) = \frac{e}{2m*c}\left(\vec{p}\cdot\vec{A} + \vec{A}\cdot\vec{p}\right) \quad (3.2)$$

with p and A being the dipole interaction and the vector potential. The Lorentz-gauge vector potential for a harmonic interaction can be written as $\vec{A}(\vec{r},t) = \vec{e}cF/2i\omega\{e^{i(\omega t-\vec{q}\cdot\vec{r})} - e^{-i(\omega t-\vec{q}\cdot\vec{r})}\}$. F is the amplitude electric field, $\vec{e}$ is the polarization vector, $\vec{q}$ is the photon wave vector at frequency ω and c is the speed of light in vacuum. Assuming that the potential vector variation is much slower than vector position ($\vec{q} \ll \vec{r} \Rightarrow e^{i\vec{q}\cdot\vec{r}} \cong 1$) and because $[\vec{A},\vec{p}] = 0$, the transition rate can be rewritten as:

$$P = -\frac{2\pi}{\hbar}\frac{e^2F^2}{4m*^2\omega^2}|\langle f|\vec{e}\cdot\vec{p}|i\rangle|^2\delta(E_f - E_i \pm \hbar\omega). \quad (3.3)$$

Writing the wavefunction of an electron in state j and band n as $\psi_j(\vec{r}) = f_j(\vec{r})u_{n,j}(\vec{r})$, the matrix element between two states can be split like:

$$\langle\psi_j|\vec{e}\cdot\vec{p}|\psi_i\rangle = \vec{e}\langle u_{mj}|\vec{p}|u_{ni}\rangle\cdot\langle f_j|f_i\rangle + \vec{e}\langle u_{mj}||u_{ni}\rangle\cdot\langle f_j|\vec{p}|f_i\rangle, \quad (3.4)$$

where i and j, and n and m are the subband and band indices of the initial and final states, respectively. The first term describes interband transitions, which are accompanied by a change of the band index n. If the initial and the final bands are the same, as in the case of intersubband transitions in the conduction band, this term vanishes. The second term, which describes transitions between subbands in the same band, becomes relevant. This term can be decomposed as the sum of a parallel and perpendicular component to the growth direction:

$$\langle f_j|\vec{e}\cdot\vec{p}|f_i\rangle = \langle e^{ik'_{\perp j}\cdot r}\chi_j|e_\perp p_\perp + e_z p_z|e^{ik_{\perp i}\cdot r}\chi_i\rangle. \quad (3.5)$$

Using $p_z = -i\hbar\,\partial/\partial z$ one obtains:

$$\langle f_j|\vec{e}\cdot\vec{p}|f_i\rangle = \hbar e_\perp k_\perp \delta_{k_{\perp i},k'_{\perp j}}\delta_{i,j} + e_z\delta_{k_{\perp i},k'_{\perp j}}\langle\chi_j|p_z|\chi_i\rangle. \quad (3.6)$$

The matrix element which determines the intersubband transition probability is proportional to:

$$\langle \chi_j | p_z | \chi_i \rangle = \int \chi_j^*(z) p_z \chi_i(z) \mathrm{d}z. \tag{3.7}$$

The operator p_z being odd, the matrix element will be not null if the two states implied in the transition are of opposite parity. The optical transition between two states can be described in terms of a dimensionless quantity known as oscillator strength instead of dipole matrix element which is defined by:

$$f_{ij} = \frac{2m_0}{\hbar^2}(E_j - E_i)\langle \psi_i | Z | \psi_j \rangle^2 = \frac{2m_0}{\hbar^2}(E_j - E_i) z_{ij}^2, \tag{3.8}$$

where z_{ij} denotes the optical dipole matrix element. The oscillator strength obeys the sum rule $\sum_j f_{ij} = m_0/m^*$ and hence, it is possible to compare different systems in term of this quantity.

3.2.1.2 Non-radiative Transitions

Scattering of electrons with impurities, interface roughness, acoustic and optical phonons are non-radiative relaxation mechanisms which are elastic processes except for the two later ones. Electron–electron scattering is another elastic scattering mechanism that an electron with higher energy transfers its energy to one or more electrons with lower energy levels and thus the total energy of the whole of electrons remains constant. The only process that allows thermalization of the electron distribution with the lattice, is the electron–phonon scattering which can be considered by the following Hamiltonian.

$$H_{\text{e-ph}} = \sum_q \left[\alpha(q) e^{(-i\vec{q}\cdot\vec{r})} a_{\vec{q}}^+ + c \cdot c. \right], \tag{3.9}$$

where $a_{\vec{q}}^\dagger$ is the creation operator for the phonon in the mode $\vec{q}$. The Fröhlich interaction strength for electron–optical phonon scattering can be defined by:

$$|\alpha(\vec{q})|^2 = 2\pi\hbar\omega_{\text{LO}} \frac{e^2}{\varepsilon_r \Omega q^2}, \tag{3.10}$$

where $\varepsilon_r^{-1} = \varepsilon_\infty^{-1} - \varepsilon_s^{-1}$ and ε_s and ε_∞ are the static and high frequency permittivities, Ω is the volume of elementary lattice cell, $E_{\text{LO}} = \hbar\omega_{\text{LO}}$ is the energy of the optical phonon. The strength of the electron–acoustic phonon interaction can be defined by:

$$|\alpha(\vec{q})|^2 = \frac{C_0}{\Omega}\hbar\omega_q. \tag{3.11}$$

In the above equations optical phonons have considered to be dispersion less while a linear dispersion is considered for acoustic phonons. The scattering rate from an initial state $|i, k_i\rangle$ to all final states $|f, k_f\rangle$ due to LO phonon emission at zero temperature can be computed from [1]:

$$\tau_i^{-1} = \frac{m^* e^2 \omega_{LO}}{2\hbar \varepsilon_r} \sum_j \int d\theta \frac{I^{ij}(Q)}{Q}, \qquad (3.12)$$

where Q is defined as:

$$Q = \left(k_i^2 + k_f^2 - 2 k_i k_f \cos\theta\right)^{1/2} \qquad (3.13)$$

$$k_f^2 = k_i^2 + \frac{2m^*}{\hbar^2}(E_i - E_f - \hbar \omega_{LO}), \qquad (3.14)$$

where E_i, E_f are the energy of initial and final state, θ is the angle between the in-plane wavevectors k_i and k_f, and $I^{ij}(Q)$ is given by:

$$I^{ij}(Q) = \int dz \int dz' \chi_i(z) \chi_j(z) \, e^{[-Q|z-z'|]} \chi_i(z') \chi_j(z'). \qquad (3.15)$$

The above expression is equal to δ_{ij} for $Q = 0$ and decays like Q^{-1} at large Q values. Important intersubband scattering mechanisms are schematically reviewed in Fig. 3.2 in a wide and narrow quantum well where the energy separation, E_{fi}, is larger or smaller than the optical phonon energy. As it is obvious, for energy separation larger than LO-phonon energy, $E_{fi} > \hbar \omega_{LO}$, electron–optical phonon scattering mechanism is the most dominant process while for energy separation smaller than LO-phonon energy, several mechanisms such as electron–electron

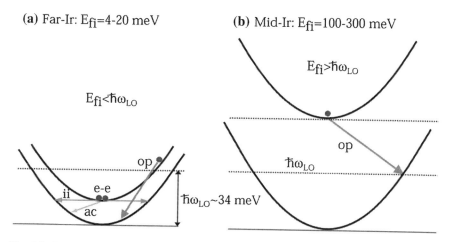

Fig. 3.2 Important intersubband scattering mechanisms, **a** the energy separation E_{fi} is smaller than LO phonon energy and electron–electron scattering (*e–e*), ionized-impurity scattering (*ii*), acoustic phonon emission (*ac*) and thermally activated LO-phonon scattering (*op*) are dominant, **b** the energy separation E_{fi} is larger than LO phonon energy and optical phonon (*op*) emissions is the dominant scattering mechanism [88]

scattering, thermally activated LO-phonon scattering and ionized-impurity scattering and electron–acoustic phonon scattering take place. The last two scatterings are specially dominant at low-temperatures in the case of a terahertz QCL where $E_{fi} < E_{LO}$. The associated lifetime for optical phonon scattering process is ~ 1 ps [1] and can reach to $\sim 0.2 - 0.3$ ps for an intersubband transition separated by $E_{fi} \approx E_{LO}$ whereas ionized-impurity and acoustic phonon scatterings are much slower and occur with a lifetime of ~ 10 ps to 1 ns [2, 3].

3.2.2 Resonant Tunneling Transport

Photon-assisted tunneling in a biased superlattice (SL) in which the ground state is lifted above the first excited of the adjacent well can lead to optical gain as it is schematically illustrated in Fig. 3.3 [4]. The states of SL are denoted by E_j and have a width δ_j much smaller than the minigaps with energy width of Δ at zero field. Then energy of mini-band j of the nth period under the applied field F can be written as:

$$E_j^n\left(\vec{k}_\perp\right) = E_j - edFn + \frac{\hbar^2 \vec{k}_\perp^2}{2m^*}, \qquad (3.16)$$

where d is the SL period as illustrated in Fig. 3.3c. By denoting the transmitted energy to the electron by the electric field for one period with edF, electrons become localized within one period and can tunnel to the next well when $edF > \delta_j$.

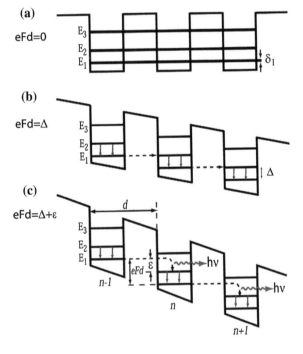

Fig. 3.3 Schematic of a superlattice structure where electrons can tunnel from the ground state of the nth well to the first excited state of the $(n + 1)$th well through emission of a photon with energy of energy $\varepsilon = eFd - (E_2 - E_1)$ [88]

3.2 Quantum Cascade Laser Principles

When $edF > E_2 - E_1 = \Delta$ as depicted in Fig. 3.3c, the first excited level in the nth well lies below the ground-state of the $(n-1)$th well.

Thus, electrons can tunnel to the excited state of the nth well by the simultaneous emission of a photon of energy $\varepsilon = eFd - (E_2 - E_1)$. Then, electrons relax non-radiatively to the ground state and become ready for the next photon-assisted tunneling.

Following non-radiative transitions from the excited to the ground state within one well the electrons are 'recycled' for another photon-assisted tunneling process. By defining Ω as the Rabi frequency, $\Omega = F\langle \phi_1^n | z | \phi_0^{n+1} \rangle$, $\tau_\perp$ as the relaxation time for the momentum in the plane of the layers and τ_2 as the lifetime of an electron in the upper laser level, the obtained current density can be expressed as [4, 5]:

$$J = edN_s \left(1 - e^{-\Delta/kT}\right) \frac{2|\Omega|^2 \tau_\perp}{1 + \varepsilon^2 \tau_\perp^2 + 4|\Omega|^2 \tau_2 \tau_\perp}, \quad (3.17)$$

where N_s is the electron sheet density and the difference between the populations of the ground and first excited state in a well is inserted to the above equation through the expression $(1 - \exp(-\Delta/kT))$. The current density will be maximum if the ground state of the $(n-1)$th well be in resonant with the excited state of the (n)th well or in the other words $\varepsilon = eFd - (E_2 - E_1) = 0$.

3.2.3 Quantum Cascade Lasers

Quantum cascade lasers consist of several repeating regions which are known as active region where population inversion takes place in a similar manner to the structure presented in Fig. 3.3 and an injector region which couples two successive active regions and enables the electrons tunneling from an active well to higher energy level in the active region of the next period [6]. This configuration consisted of two active regions and an injection region (one and half period) is schematically drawn in Fig. 3.4.

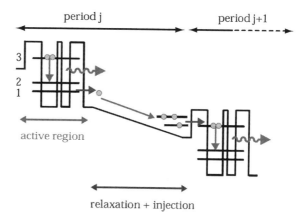

Fig. 3.4 Schematic one and half period of a quantum cascade laser consisted of two active regions where lasing take place and an injector region which injects the electrons into the upper laser energy state (level 3) of the subsequent active section [88]

3.2.4 Optical Gain

In order to obtain the optical gain of a QCL, the carrier dynamics inside gain (active) region can be considered through a simplified three-level model as shown in Fig. 3.5. In this model, electrons are injected into level $|3\rangle$ with a current density of J and an injection efficiency of η and the lasing transition takes place between levels $|3\rangle$ and $|2\rangle$. The injected electrons to level $|3\rangle$ may relax to level $|2\rangle$ and $|1\rangle$ and either escape to the injector through tunneling in the continuum with a total rate of $\tau_3^{-1} = \tau_{32}^{-1} + \tau_{32}^{-1} + \tau_{esc}^{-1}$. The rate equations of the system including the sheet densities n_2 and n_3 and the photon density S can be described by:

$$\frac{dn_3}{dt} = \eta \frac{J}{q} - g_c S(n_3 - n_2) - \frac{n_3}{\tau_3} \tag{3.18}$$

$$\frac{dn_2}{dt} = (1-\eta)\frac{J}{q} + \frac{n_3}{\tau_{32}} + g_c S(n_3 - n_2) - \frac{n_2}{\tau_2} \tag{3.19}$$

$$\frac{dS}{dt} = \frac{c}{n}\{[g_c(n_3 - n_2) - \alpha]\}S + \beta\frac{n_3}{\tau_{sp}}, \tag{3.20}$$

where τ_2 is the depopulation time of the level 2. Since the spontaneous emission lifetime at terahertz frequencies is much longer than non-radiative lifetimes, radiative relaxation has no effect in transport below threshold. Also, g_c is the gain cross section, c is the speed of light in vacuum, n is the mode refractive index and α is the total optical loss, $\alpha = \alpha_w + \alpha_m$, including waveguide loss α_w and mirror loss α_m.

In order to achieve the population inversion ($n_3 - n_2$) one can solve stationary rate equations ($dn/dt = 0$) which yields:

$$\Delta n = n_3 - n_2 = \frac{\frac{J}{q}\left(\eta + \frac{\tau_2}{\tau_3}(1-\eta) - \frac{\tau_2}{\tau_{32}}\right)}{\frac{1}{\tau_3} + g_c S\left[1 + \tau_2\left(\frac{1}{\tau_3} - \frac{1}{\tau_{32}}\right)\right]}. \tag{3.21}$$

To insure the gain condition in the three level model of Fig. 3.5 the equilibrium inversion at the absence of optical field ($S = 0$) should be obtained from the above equation which yields (the injection efficiency is assumed to be unity or $\eta = 1$):

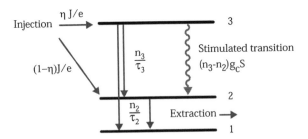

Fig. 3.5 Three level model of an active region including carrier population dynamics. Lasing transition occurs between levels 3 and 2 and carriers are extracted through sequential tunneling to the injector region [88]

3.2 Quantum Cascade Laser Principles

$$\Delta n = n_3 - n_2 = \frac{J}{q}\left(1 - \frac{\tau_2}{\tau_{32}}\right)\tau_3. \quad (3.22)$$

Therefore, population inversion requires $\tau_{32} > \tau_2$ meaning that the carriers of lower laser level (level $|2\rangle$) must be extracted faster than the non-radiative repopulation of this level from level $|3\rangle$.

3.2.5 Threshold Current

By solving Eq. 3.20 stationary and neglecting the spontaneous emission he threshold current density, J_{th}, is obtained as:

$$J_{\text{th}} = \tau_3 \frac{1}{q} \frac{\alpha/g_c}{1 - \tau_2/\tau_{32}}, \quad (3.23)$$

where

$$g_c = \frac{4\pi q^2}{\varepsilon_0 n \lambda} \frac{z_{32}^2}{2v_{32}L_p}. \quad (3.24)$$

The gain coefficient of the QCL can then be obtained using Eq. 3.22 as:

$$g = \tau_3 \left(1 - \frac{\tau_2}{\tau_{32}}\right) \frac{4\pi q}{\varepsilon_0 n_{\text{eff}} \lambda} \frac{z_{32}^2}{2v_{32}L_p}. \quad (3.25)$$

In above equation λ is the emitted wavelength, z_{32} is the dipole matrix element between states $|3\rangle$ and $|2\rangle$, n_{eff} is the effective mode refractive index, L_p is length of one period of QCL including an active and an injection regions and $2v_{32}$ is the full-width at half-maximum (FWHM) of the intersubband electroluminescence.

3.2.6 Losses

The propagating mode insight the laser waveguide experiences three major loss mechanisms. The first loss mechanism arises from the cavity facets which form the laser resonator. These facets are typically un-coated, as-cleaved, and parallel semiconductors which provide a reflectivity that can be approximate by $R = (n_{\text{eff}} - 1)^2/(n_{\text{eff}} + 1)^2$. Incompleteness of these mirrors is source of the loss expressed by:

$$\alpha_m = -\frac{1}{2L}\ln(R_1 R_2), \quad (3.26)$$

where L is the length of the resonator and R_1 and R_2 are the reflectivity of the mirrors. Free-carrier absorption is the second source of loss which specially becomes important in the doped semiconductor regions and the metallic contact layers. This loss mechanism is also known as waveguide loss denoted by α_w. The third possible loss source is related to carrier absorption in the injector region due to high dipole matrix element of inter-miniband transitions. Thus, any optical transition in the injector region with the same wavelength of the laser emission will cause this type of loss. According to the mentioned loss mechanisms, the threshold current density can be described in the form:

$$g\Gamma J_{th} = \alpha_w + \alpha_m, \qquad (3.27)$$

where Γ is the confinement factor which determines the fraction of the optical mode that overlaps with the active region of the laser and the parameter $g\Gamma$ is known as the modal gain.

3.2.7 Slope Efficiency

The slope efficiency is defined as the increase in optical power (per outcoupling facet) per unit current, given by:

$$\frac{\partial P}{\partial I} = N_p h\nu \alpha_m \frac{\partial S}{\partial J} = N_p \frac{h\nu}{q} \frac{\alpha_m}{\alpha_w + \alpha_m} \frac{\tau_{eff}}{\tau_{eff} + \tau_2}. \qquad (3.28)$$

In extraction of above equation the injection efficiency, η, is considered to be unity. Also, P is the output optical power, I is the injection current, $h\nu$ is the photon energy, $\tau_{eff} = \tau_3(1 - \tau_2/\tau_{32})$ is the effective lifetime. The slope efficiency value has direct relation with N_p which is the number of cascade periods and hence increasing the number of periods will result in increased output power.

3.3 Terahertz Quantum Cascade Lasers

After invention of terahertz QCL [7] several issues were determined as the main challenges toward realization of these kinds of terahertz sources. Efficient population inversion at small transition energies related to terahertz frequencies and designing low-loss waveguides for proper confinement of optical mode have been the main challenges in this way.

In a terahertz QCL since the intersubband transition energy is lower than LO-phonon energy, therefore, the emission probability of polar optical phonons is reduced considerably which results in increasing the upper laser lifetime. This condition is suitable for efficient population inversion, however, other non-radiative mechanisms such as electron–electron scattering, electron–impurity scattering, or electron–acoustic phonon scattering limits the operation performances. Also, it is possible to suppress polar optical phonon emission by lowering the operation

3.3 Terahertz Quantum Cascade Lasers

temperature whereas this is not easily applicable in the case of electron–electron scattering since this process depends on several parameters such as doping sheet density and multiple space covering. Exploiting the benefits of surface plasmons in realization of terahertz waveguides has also lead to development of low-loss waveguides, $\alpha_w = 5\text{--}10 \text{ cm}^{-1}$, with high confinement factors ($\sim 30\%$) [7–9].

3.3.1 Terahertz QCL Structures

Due to special active region transition energy requirements of terahertz QCL, four different designs (chirped SL, bound-to-continuum, and resonant phonon) have been proposed for the active region of these lasers since their first demonstration in 2001. These designs are schematically illustrated in Fig. 3.6 [10].

3.3.1.1 Chirped-Superlattice

The Chirped SL QCLs [7, 11, 12] are one of the promising designs for operation in long wavelengths that lasing transition occurs between two minibands rather than between subbands of an active well. By varying the thickness of barriers and wells in this type of active regions, it is possible to control the width of minibands and minigaps respectively. For the terahertz QCLs with miniband width lower than LO phonon energy, electrons are extracted from the lowest miniband through electron–electron scattering, electron–impurity scattering, and interface roughness scattering and are injected into the upper miniband of the following stage. This design reduces the probability of non-radiative scattering from the upper laser level into the lower laser level. Also, due to large spatial extent and overlap of the wavefunctions, high oscillator strength and gain is expected in this kind of active regions [13]. A schematic of SL QCL is presented in Fig. 3.6 where radiative transition has designed to take place from the lowest state of the upper miniband '2' to the top state of the lower miniband '1'.

Pulsed and continuous-wave SL QCLs at 3.5 THz with low-threshold-current densities (0.095 and 0.115 kA/cm^2 respectively) are demonstrated examples of this type of lasers [14]. Despite of large dipole transition matrix element in the chirped quantum wells and barriers, the thermal backfilling of the lower miniband and consequently a weak population inversion lead to limitation in slope efficiency and maximum operation temperature in SL QCLs.

3.3.1.2 Bound-to-Continuum Transition

An alternative approach for SL QCLs is realization of terahertz transitions based on a bound-to-continuum transition [15–17]. The emission process happens between an isolated state created inside a minigap by a thin well adjacent to the

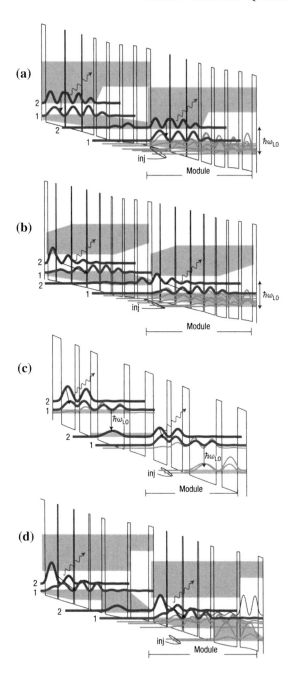

Fig. 3.6 Schematic of four main designs of QCL active regions. **a** Chirped superlattice, **b** bound to continuum, **c** resonant–phonon and **d** hybrid/interlaced designs [10]

injection barrier and the lower miniband. Similar to chirped SL structure, the depopulation process of carriers is mediated via electron–electron scattering and resonant tunneling inside the lower miniband. Since the lasing transition from the upper level to lower miniband is diagonal in real space, the oscillator strength of the transition decreases slightly as the overlap with the miniband states decreases, but the upper-state lifetime increases as non-radiative scattering is similarly reduced and the injection efficiency is increased since the injector states couple strongly to upper lasing state compared with lower miniband. In fact, this type of active region combines long lifetime and good injection efficiency of the upper state with high extraction efficiency of lower miniband where both of the properties are essential for efficient population inversion in terahertz QCLs. The combination of these properties yield improved temperature and power performance in comparison with the chirped SL design. Terahertz lasers with pulsed operation up to 100 K at 3.5 THz and CW powers up to 50 mW at 10 K have been demonstrated based on this kind of active regions [17].

3.3.1.3 Resonant Phonon Structure

The third type of active regions introduced for realization of terahertz QCLs utilizes combination of resonant tunneling and LO phonon scattering processes for depopulation instead of using SL [18–20]. Depopulation of carriers through LO phonon emission is a fast and robust process which is not highly temperature dependant and on the other hand, the energy separation between the lower laser state and injector level is determined by LO phonon energy ($\geq E_{LO} = 36$ meV) which limits the thermal backfilling of the lower laser state. Also, it is possible to utilize the benefits of both bound-to-continuum design and resonant phonon depopulation method to achieve higher performances. A pulsed mode terahertz QCL utilizing this approach with maximum operating temperature of 116 K is reported in [21].

3.3.1.4 Hybrid/Interlaced Design

In hybrid design of active regions phonon-assisted depopulation has been incorporated with a bound-to-continuum transition as illustrated in Fig. 3.6d. These structures are also known as 'interlaced' structures, because photon- and phonon-emission events take place alternatively. This kind of active region design is particularly interesting for achieving very long wavelength operation.

3.3.1.5 Waveguide Design

A low-loss waveguide resonator along with high optical confinement factor is one of the essential requirements of terahertz QCLs. The common resonator structure of QCLs is a ridge waveguide where light propagates along the ridge region and

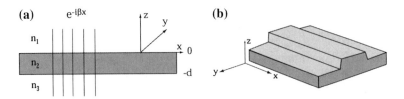

Fig. 3.7 a Typical waveguide geometry with a core layer with refractive index of n_2 and two cladding layers with refractive indices of n_1 and n_3 where the optical mode is confined inside the core layer if $n_2 > n_1, n_3$, and **b** schematic view of a ridge waveguide structure [88]

reflects from the mirrors at the front and end of the structure. A schematic view of a ridge waveguide is illustrated in Fig. 3.7 where the ridge region provides mode confinement in the y-direction.

In order to investigate the waveguide specification required for QCLs, here we start the discussion with an overview on waveguide structure of mid-infrared QCLs then the subject will be generalized to longer wavelengths.

Dielectric waveguides have been used to provide optical confinement in the growth direction (z-direction) by assuring that the refractive index of the active region is higher than the cladding regions as illustrated in Fig. 3.7. These waveguides can be modeled by a slab structure with infinite extension in x- and y-directions and a finite extension in z-direction. The wave equation for such a structure can be written as:

$$\nabla^2 \vec{E}(\vec{r}) + k^2 n^2(\vec{r}) \vec{E} = 0, \quad (3.29)$$

where $\vec{E}$ is the optical electric field, $n(\vec{r})$ is the refractive index, $\vec{r} = (x, y, z)$ and $k = \omega/c$, with c is the speed of light in vacuum. The solution of the wave equation gives the optical mode traveling in x-direction with a propagation constant β in the form of:

$$\vec{E}(\vec{r}, t) = E(y, z) e^{[i(\omega t - \beta x)]} \quad (3.30)$$

by replacing this solution in the wave equation one obtains:

$$\left(\frac{\partial^2}{\partial y^2} + \frac{\partial^2}{\partial z^2} \right) \vec{E}(y, z) + \left[k^2 n^2(\vec{r}) - \beta^2 \right] \vec{E}(y, z) = 0. \quad (3.31)$$

Considering a planar model where no variation exists in y-direction ($\partial/\partial y = 0$) Eq. 3.31 reduces to:

$$\frac{\partial^2}{\partial z^2} E(y, z) + \left[k^2 n_j^2(\vec{r}) - \beta^2 \right] E(y, z) = 0, \quad (3.32)$$

where $j = 1, 2, 3$ denote different regions in the slab waveguide as displayed in Fig. 3.7a ($j = 1$, if $x > 0$, $j = 2$ if $-d < x < 0$ and $j = 3$ if $x < -d$). The condition on refractive indices, $n_2 > n_1, n_3$, imposes a condition on the propagation constant β which should satisfy $kn_1 < kn_3 < \beta < kn_2$ in order to have guided wave. For TE modes of the waveguide in the symmetric case ($n_1 = n_3$), the solution of Eq. 3.32 can be expressed as [22]:

$$E_y(z) = \begin{cases} A\exp(-kz) & \text{for } z \geq 0 \\ B\cos(\alpha z) + C\sin(\alpha z) & \text{for } -d \leq z \leq 0 \\ D\exp(k(z+d)) & \text{for } z \leq -d. \end{cases} \quad (3.33)$$

Applying the continuity conditions for obtained solutions and their derivates ($\partial E_y(z)/\partial z = 0$) at both $z = 0$ and $z = -d$, yields:

$$\begin{cases} \tan(\alpha d) = \frac{2k\alpha}{\alpha^2 - k^2} \\ \beta^2 + \alpha^2 = n_1^2 k^2 \\ \beta^2 - k^2 = n_2^2 k^2 \\ k = \omega/c \end{cases} \quad (3.34)$$

For given values of ω and d, a propagation constant which satisfy the above condition is the propagation constant of the guided mode of the waveguide and by varying the radial frequency, the dispersion characteristics of the waveguide (i.e. β (ω) or n_{eff} (λ)) is obtained where n_{eff} is the refractive index which is defined by $n_{\text{eff}} = \beta/k = \beta c/\omega$. Depending on the whether the waveguide is lossless or not, the refractive index n_{eff} can be real or complex. In a waveguide with complex refractive index, $n_{\text{eff}} = n + ik$, $k > 0$ attributes to loss while $k < 0$ corresponds to gain.

InP-based material systems are the most common materials for realization of mid-infrared QCL waveguides since they can provide considerable refractive index differences with InP ($n_{\text{InP}} \cong 3.10$) and AlInAs ($n_{\text{AlInAs}} \cong 3.20$) cladding layers and InGaAs/AlInAs core layers (($n_{\text{InGaAs}} \cong 3.49$)). A linear interpolation is typically used to determine the effective refractive index of the active region consisted of two or more different layers type according to their volume fraction. Although the design process of mid-infrared QCLs seems to be an easy task, one should consider the effect of the doped waveguide layers in these lasers for carrier transport during current injection mechanism. Figure 3.8 shows mode intensity and refractive indices of a typical waveguide of QCLs which presents the doping levels of different layers in the inset of the figure.

Free carrier absorption is a loss source in doped layers which enhances the waveguide loss. Also, coupling of the mode with contact metal is another waveguide loss source which can be solved by adding a thin and high-doped InGaAs layer between the contact and the semiconductor.

GaAs/AlGaAs material system is another alternative that have been considered for development of QCLs, however, due to highest refractive index of GaAs substrate, it cannot serve as cladding layer for QCLs developed based on this material system.

On the other hand, using GaAs/AlGaAs multilayer as core and AlGaAs as cladding layers requires a high-AlAs-mole-fraction material whereas high-doping of these materials for efficient current transport is a difficult task. As a solution for this problem, researches on plasmon-enhanced dielectric waveguides with highly

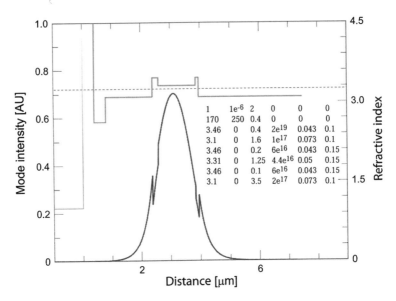

Fig. 3.8 Mode intensity and refractive indices for a dielectric waveguide of QCL for a laser wavelength of 3.5 μm [88]

doped GaAs for both claddings have shown promising results for GaAs/AlGaAs QCLs [23–27].

Waveguide fabrication for QCLs at longer wavelengths encounters two main challenges: the thickness of cladding layer (at least in the order of $\lambda/2$) becomes incompatible with MBE technology and due to increased growth time, the number of defects in the structure is increased. Also, for frequencies above the plasma frequency, free carrier absorption loss increases with λ^2. Surface-plasmon waveguides [28–30] where the modes propagate in the metal–semiconductor interface (two media with opposite dielectric constants in sign) are the solution of these challenges which do not need additional confinement layers since the optical mode exponentially decays in the two directions normal to the interface of metal–semiconductor. In surface-plasmon waveguide, the metal layer is deposited directly above the active region of the laser and serves as a guiding interface. However, due to mode penetration into the metal medium, electromagnetic surface waves experience high optical losses in a metal-dielectric interface which can be described by an attenuation coefficient α given by [31]:

$$\alpha \cong \frac{4\pi n n_d^3}{\kappa^3 \lambda}. \tag{3.35}$$

The real and imaginary parts of the metal complex refractive index are represented by n and κ in the above equation. Also, n_d is the refractive index of the dielectric, and λ is the wavelength in vacuum. The interface losses is inversely proportional to λ and it can be minimized by selecting a metal that meets $\kappa \gg n$, in the other words, a metal with real and strongly negative dielectric constant.

Higher confinement factor of active region compared to conventional multi-layer waveguides and strong coupling to the active material are the benefits of these types of waveguides whereas the increased absorption losses should be traded off with the mentioned advantageous. Long-wavelength QCLs have been realized using this method reported in [32–34].

3.3.2 Drude Model

The Lorentz–Drude theory is a simple phenomenological approach based on the classical dispersion theory which can be used in order to obtain the effect of conductivity on the complex permittivity for the semiconductor and metal layers that result from free carrier effects. In Drude model, conduction electrons are assumed to move freely in the presence of an applied field $E(t) = \text{Re}E(\omega)e^{-i\omega t}$, subject only to a collisional damping force. The probability of a collision during an interval dt is dt/τ, where τ is the scattering time, or equivalently, the average time between collisions. Frequency dependant conductivity in Drude model is expressed by:

$$\sigma = \frac{ne^2\tau}{m^*(1 - i\omega\tau)}, \tag{3.36}$$

where n is the carrier density, τ is the critical scattering time, ω is the light pulsation, and m^* is the effective carrier mass. By defining the frequency dependent permittivity with $\varepsilon(\omega) = \varepsilon_{\text{core}} + i\sigma/\tau$, the dielectric constant which describes the optical properties of media can be expressed as:

$$\varepsilon(\omega) = \varepsilon_{\text{core}} + i\frac{\sigma}{\tau} = \varepsilon_{\text{core}}\left[1 - \frac{\omega_p^2\tau^2}{1+(\omega\tau)^2} + i\frac{\omega_p^2\tau}{\omega\left(1+(\omega\tau)^2\right)}\right], \tag{3.37}$$

where $\omega_p^2 = ne^2/m*\varepsilon_{\text{core}}$ is the plasma frequency of the material and $\varepsilon_{\text{core}}$ is the core permittivity. If $\text{Re}(\varepsilon) > 0$ the semiconductor serves as a dielectric medium and for s behaves as plasma by the proper choice of carrier concentration. In the above equation, if $\omega\tau \gg 1$ the conductivity, $\sigma = ine^2/m^*\omega$, is purely imaginary and the dielectric permittivity reduces to the form:

$$\varepsilon(\omega) = \varepsilon_{\text{core}}\left[1 - \frac{\omega_p^2}{\omega^2} + i\frac{\omega_p^2}{\omega^3\tau}\right]. \tag{3.38}$$

For radiation frequencies lower than the plasma frequency, $\omega < \omega_p$, the permittivity is negative, $\varepsilon < 0$, and the refractive index is purely imaginary ($\bar{n} = i\kappa$). For radiation frequencies much larger than the plasma frequency, $\omega \gg \omega_p$, the factor $1 - \omega_p^2/\omega^2$ in above equation is reduced and the free carrier loss can be expressed by:

$$\alpha_{fc} = \frac{\omega_p^2}{\omega^2} \frac{1}{c\tau} \sqrt{\frac{\varepsilon}{\varepsilon_0}} = \frac{ne^2 \lambda^2}{4\pi^2 m^* c^3 \tau \varepsilon_0} \sqrt{\frac{\varepsilon}{\varepsilon_0}} \propto \frac{\lambda^2}{\tau}. \quad (3.39)$$

3.3.3 Terahertz-Waveguide

Although single surface-plasmon waveguide have exhibited high performance for long wavelength QCLs (typical range of $\lambda = 17$–24 μm), the situation is completely different for far-infrared QCL waveguides ($\lambda \geq 50$ μm). Specially, at terahertz frequencies the waveguides should be based on a metal–semiconductor–metal geometry. Due to fabrication difficulties, the lower metal layer is usually replaced with a highly $n{++}$ semiconductor layer with a thickness of ~ 1–2 μm. The difference between the mode confinement factors of single and double surface-plasmon (metal–semiconductor–metal structure) waveguides is illustrated in Fig. 3.9a, b respectively.

Despite of near unity confinement factor in Fig. 3.9b, this is achieved at the hands of high propagation loss which makes the demonstration of terahertz QCLs a challenging fact. The optical losses of double surface-plasmon waveguide at terahertz frequencies have measured to be $\alpha_w = 42 \pm 20$ cm^{-1} at $\lambda = 75$ μm mainly due to free carrier absorption in the lower cladding layer and doped substrate due to the significant skin depth [35]. The waveguide structure and typical two-dimensional mode intensity pattern of this kind of waveguides are presented in Fig. 3.10 where the remaining ~ 10-μm-thick epitaxial active region after wafer-bonding and substrate removal, is patterned by photolithography and is

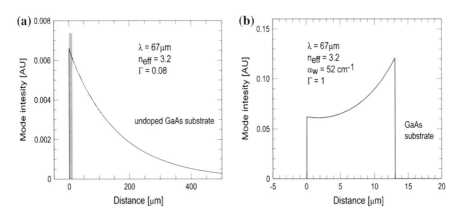

Fig. 3.9 a Mode intensity of a 67 μm mode in (a) single surface-plasmon waveguide with a undoped GaAs substrate which results in long decay process in active region and small confinement factor, 0.08, is obtained. b Double surface-plasmon waveguide where the active region is limited to a gold layer at the left and a high-doped, $n = 5 \times 10^{18}$ cm^{-3}, GaAs layer grown on a $n + $ GaAs substrate at the right side which results in ~ 1 confinement factor [89]

3.3 Terahertz Quantum Cascade Lasers

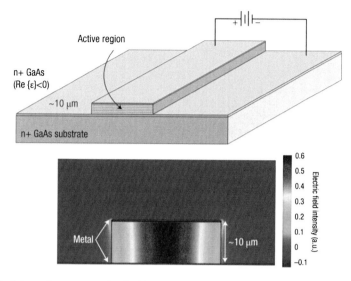

Fig. 3.10 Schematic structure of a double-surface plasmon terahertz QCL waveguide with two-dimensional mode intensity pattern [10]

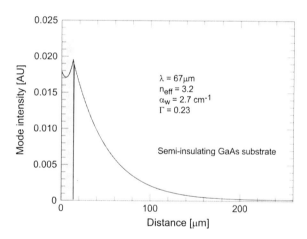

Fig. 3.11 Mode intensity of semi-insulating GaAs-based single surface-plasmon QCL waveguide at $\lambda = 67$ µm [88]

etched (typically) into ridges to produce a structure similar to a microstrip transmission line.

An alternative solution have proposed based on semi-insulating substrates where the modes are bound to the upper metallic contact and a thin heavily doped contact layer grown directly beneath the active region and above the semi-insulating GaAs substrate [7, 8]. The mode penetrates into the substrate because the doped layer is thinner than its skin depth and the free carrier absorption becomes minimized due to small overlap of the mode and the doped contact layer. Figure 3.11 presents the mode intensity in semi-insulating GaAs-based waveguide structure at $\lambda = 67$ µm where the waveguide losses have decreased to 2.7 cm^{-1}.

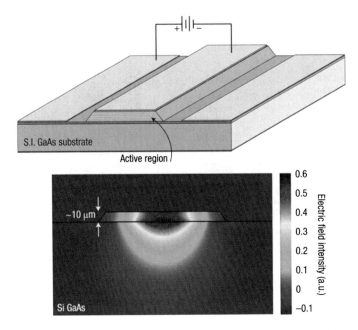

Fig. 3.12 Schematic structure of a semi-insulating surface-plasmon terahertz QCL waveguide with two-dimensional mode intensity pattern [10]

Proper doping level of the lower contact layer insures the creation of surface-plasmons between the contact layer and the semi-insulating substrate (negative dielectric constant for lower contact layer). The waveguide structure and two-dimensional mode intensity pattern of this type of waveguides are presented in Fig. 3.12 which involves the growth of a thin (0.2–0.8 μm thick) heavily doped layer underneath the 10-μm-thick active region, but on top of a semi-insulating GaAs substrate.

3.3.4 Distributed Feedback QCLs

The introduced waveguides for QCLs up to now have been based on Fabry–Perot (FP) cavities which exhibit a relatively broadband and multimode operation suitable for a specific class of spectroscopy applications such as targeting liquids and solids. However, applications such as gas sensing, remote chemical sensing and pollution monitoring need a tunable source with a narrow linewidth. To achieve single-mode operation of QCLs, Distributed feedback (DFB) QC structures have been introduced in mid-infrared region which have provided continuously tunable single-mode output [36–41]. Since the top cladding of QCLs consists of several layers (which differ in doping levels), it is possible to produce a wavelength selective feedback through grating modulation. Incorporating a grating with period Λ into the waveguide reduces the out-coupling loss for a specific wavelength (close to Bragg wavelength $\lambda_B = 2n_{eff} \Lambda$) and thus lowers the threshold gain.

In the following the two fundamental methods for demonstration of single-mode QCL based on DFB waveguides (loss coupled [36] and Index coupled QC-DFB lasers [37]) are introduced.

3.3.4.1 Loss-Coupled QC-DFB Lasers

In the Loss-coupled DFB QCLs, the highly doped topmost waveguide layer is etched and the grating pattern is overlaid into the surface of the waveguide. Etching the grating, results in loss and refractive index modulation in the waveguide. The former rises from the fact that the waveguide loss is higher in the grating grooves than the non-etched parts of the grating which cause a loss modulation while the latter happens due to the metal layer in the grating grooves which pulls the mode toward itself and modulates its effective refractive index. Because of these two modulation processes, DFB QCLs are called complex-coupled structures. A Bragg grating composed of two alternating metal stripes (e.g. gold and gold-over-titanium [34]) can affect the waveguide effective refractive index (due to metal's dielectric function) and result in single-mode operation at far-infrared wavelengths. The amount of coupling in cavity between the forward and backward travelling waves can determine the grating strength approximated by the coupling coefficient given by:

$$\kappa = \frac{\pi}{\lambda_B} n_1 + i\frac{\alpha_1}{2}, \tag{3.40}$$

where n_1 is the amplitude of the periodic modulation of the real part of the effective index (n_{eff}) of the mode, induced by the grating of periodicity Λ and α_1 is the amplitude of corresponding modulation of the absorption coefficient. For optimum performance in slope efficiency and threshold current, the product $\kappa L_{cav} \approx 3\text{--}5$, must be kept close to unity (L_{cav} is the cavity length).

In spite of easy fabrication and time-saving processing technology required for realization of this kind of etching, its distance from the active region of the laser (≥ 2 μm) reduces the mode coupling. Decreasing the thickness of upper cladding layer may provide strong coupling but this will lead to increased waveguide loss due to the absorption of the top metal layer which decrease the DFB laser performance.

3.3.4.2 Index Coupled QC Lasers

To achieve a higher performance of DFB QCLs, one can position the grating close to the active waveguide core, where the mode intensity is high and allows strong coupling with negligible additional loss [37, 42]. In order to demonstrate this type of QCL, at first the active region section is grown embedded between two layers (e. g. InGaAs) where the upper layer serves as the host region for the grating. Then the first-order grating is transferred by contact lithography and wet chemical etching. A top cladding layer (e. g. InP) is grown on top of the Bragg grating using solid-source MBE. The grating depth and duty cycle during the etching process and the reflow of material in the regrowth process allow controlling the grating strength.

3.4 Analysis of Transport Properties of THz QCLs

As it is a known fact, mid-infrared and THz QCLs operate on the same principle based on intersubband transition. However, due to different amount of subband separation which is higher than LO-phonon energy in mid-infrared QCLs and smaller than $\hbar\omega_{LO}$ in THz QCLs, the electron transport dynamics id qualitatively differs in these two types of QCLs. In the case of mid-infrared QCLs, the electron transport is dominated by LO-phonon scattering while in THz QCLs, only the high-energy tail of a hot electron distribution is subject to the LO-phonon scattering, which results in a significantly higher temperature sensitivity for the electron transport and a far greater importance of electron–electron scattering. This complicated transport mechanism makes the population inversion in THz QCLs a difficult affair. Modeling this transport mechanism may help to extend the operation of THz QCLs to higher temperatures and broader frequency ranges. Between the different modeling methods for simulation of THz QCLs such as popular rate-equation methods, Monte Carlo (MC) simulations are an efficient method for covering temperature and density-dependent scattering processes [43–46]. A flow chart of MC simulation steps is presented in Fig. 3.13 which involves electron–electron and electron–phonon interactions in one module of a THz QCL.

Figure 3.15 illustrates the MC simulation results of a 3.4 THz (87 μm or 13.9 meV) QCL which is presented with the conduction band profile in Fig. 3.14. The laser consists of a four-well structure as one period has illustrated inside the dashed box in Fig. 3.13 and the total number of periods is equal to $N_p = 175$. This structure utilizes LO-phonon scattering for depopulation of the lower state which is highly selective, as only the lower lasing level (i.e. level 4) is at resonance with a level 3 in the adjacent well. The radiative transition which is spatially vertical, takes place between levels 5 and 4 and hence, a considerable oscillator strength ($f = 0.96$) is obtained.

The calculated and measured I–V characteristics are presented in Fig. 3.15a. It is obvious that considering the effect of electron–impurity scattering, highly influences the peak current density. However, since the whole MQW structure is treated as a single quantum mechanical system in the MC studies, coherent interaction and time evolution is ignored and transport is considered as intersubband scattering among the spatially extended subband states, similar to the Boltzmann transport equation. Hence, the thickness of potential barriers does not affect the injection of electrons to the upper and their removal from the lower radiative levels and these processes are quiet efficient at resonance. While, the dephasing scattering due to interface roughness and impurity scattering interrupts the coherent interactions between states in real devices and effectively localizes

3.4 Analysis of Transport Properties of THz QCLs

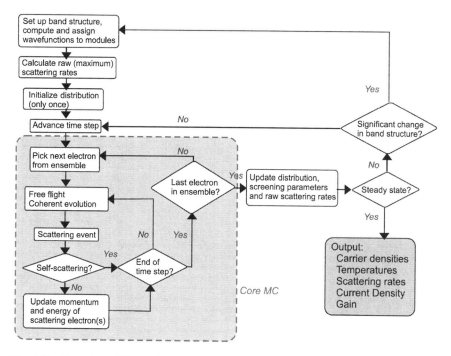

Fig. 3.13 Flow chart of Monte Carlo simulation steps for a THz QCL [46]

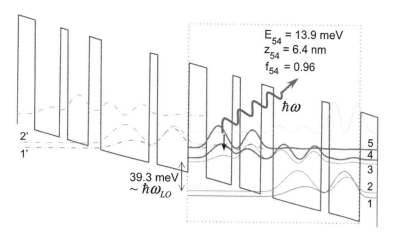

Fig. 3.14 Conduction band profile of a THz QCL biased at 64 mV/period. The layer thickness sequences beginning with the injector barrier in Angstrom are: 54, 78, 24, 64, 38, 148, 24 and 94 A respectively [44, 45]

wavefunctions which leads to an incoherent tunneling process between weakly coupled states. This incoherent tunneling process has less efficiency than the rates obtained from the Boltzmann-like model for electron injection and removal.

Fig. 3.15 **a** Current density, **b** electron temperature for subbands involved in the radiative transition, ($n = 4$ and $n = 5$), **c** subband population density for $n = 4$ and $n = 5$, and **d** material gain with threshold gain values for an uncoated-facet structure and a one-facet high reflection-coated structure as a function of bias current. The measured current density is represented with *solid curve* in panel (**a**). Also, *circle* symbols stand for the simulation results without including electron–impurity scattering and *diamond* symbols stand for the results in the presence of electron–impurity scattering [46]

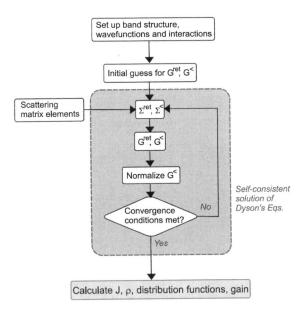

Fig. 3.16 Transport analysis flow chart of nonequilibrium Green's function approach [46]

To deal with this problem, different possible methods such as tight-binding model in density-matrix formalism and nonequilibrium Green's function approach can be utilized.

3.4 Analysis of Transport Properties of THz QCLs   215

The first method, the Rabi oscillation between two states across an energy barrier is damped by the dephasing scattering which results in reduction of the coupling among spatially localized basis states.

In the nonequilibrium Green's function approach which has introduced through the flow chart of Fig. 3.16 for the presented THz QCL structure, all the dephasing processes are considered and the important parameters like the gain and current density are calculated. Although this method is highly computation intensive, comparing the obtained results from this approach with experimental results declares its exactness compared with the results obtained from semi-classical Boltzmann approach.

3.5 High Power QCLs

The output power level of room temperature CW midinfrared QCLs has increased by orders of magnitude since their first demonstration [47–49]. Thanks to room temperature high power QCLs, applications such as photo-acoustic spectroscopy [50], remote sensing [51], infrared countermeasures and free space communication [52], have been developed in recent years. Room temperature CW operation of QCLs with considerable enhancement in wall plug efficiency (WPE) as a measure of device quality (the ratio of the radiant flux i.e. the total radiometric optical output power of the device, measured in watts, and the electrical input power i.e. the efficiency of converting electrical to optical power) and high output power of 2.5 W at 4.8 μm have been demonstrated [53] due to optimization of strain-balanced SL growth, combined with advanced thermal design, fabrication and packaging.

The number of periods, N, in a QCL for pulsed mode operation, determines several important optical properties, such as confinement factor, Γ, and free carrier waveguide loss (both decrease with N) which leads to a lower threshold current density and higher slope efficiency, as long as spatial hole burning effects are minimal. But, since the operating voltage scales with N, the WPE increases only slowly and growing the QCL core with the largest possible N yields the highest pulsed WPE. However, a thicker QCL core degrades the thermal conductance of the device and influences the CW operation. Narrower QCL waveguide and buried ridge geometry can improve the thermal conductance to achieve both the pulsed and CW output power and efficiency.

The temperature dependent behaviors of the mentioned high output power QCL at 4.8 μm in the pulsed mode operation are presented in Fig. 3.17. The temperature dependent power performances can be understood from the power–current (P–I) curves of Fig. 3.17a. The rollover current (the current at which the output power saturates) increases monotonically as a function of temperature due to the decrease of the upper laser level lifetime and the increase of the ionization of the dopants as the temperature increases. The temperature dependent current–voltage (I–V) behavior also demonstrates an increase of the operating voltage at low temperatures which can be relates to several factors. At lower temperatures, more voltage is

Fig. 3.17 a Power–current–voltage and **b** WPE-current curves of the high power QCL operating in pulsed mode at temperatures between 80 and 298 K. The cavity length and ridge width are 5 mm and 8.6 μm, respectively [54]

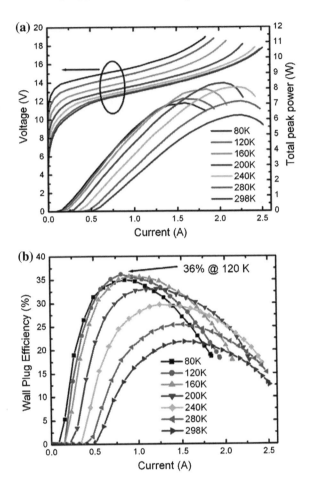

required to populate the upper laser level since most of the carriers are located at the bottom of the injector. The conduction band offset of the system is slightly higher resulting in a decrease of wavelength and a higher operating voltage and the parasitic voltage drop such as the contact voltage drop and the junction voltage drop at the interfaces of two materials with abrupt doping changes gets more pronounced.

Figure 3.17b illustrates the current dependent WPE at different temperatures where both the global WPE of 36% and the room temperature WPE of 22% are the highest values reported to date for all QCLs.

Figure 3.18a displays the corresponding P–I and I–V curves which denotes a maximum output power of 3.4 W at 1.75 A. The appeared kink around 1.3 A in the P–I curve, in which the WPE reaches its maximum as illustrated in Fig. 3.18b, is also present at higher temperatures in CW operation, but absent in the pulsed mode operation can be attributed to a signature of thermally induced mode instability. Right at the kink, the CW WPE reaches a record value of 16.5%, with 15.5% still observed in the vicinity of the kink.

3.5 High Power QCLs

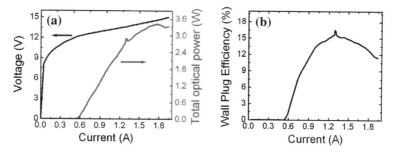

Fig. 3.18 a Power–current–voltage, and b WPE-current curves of the high-power QCL operating in CW mode at room temperature [54]

For long wavelength QCLs ($\lambda \geq 10$ μm), similar successes for high-power operation at the wavelength of 10.2 μm at room temperature with 2.2 W average power and 0.62 W CW power [54] compared with previously reported structures with average and CW output powers of 310 mW at 10.5 μm [55] and 148 mW at 10.6 μm [56] has obtained.

Increasing waveguide loss (approximately as λ^2) and decreasing optical confinement are the main challenges for long wavelength QCLs which can be partially overcome by increasing the number of QCL periods. Figure 3.19 depicts the low duty cycle (1%) operation of a 95 μm wide, 5 mm long, uncoated QCL structure where the heat sink temperature was changed from 298 to 373 K. Since the laser core is much thicker and wider than a typical QCL in the short wavelength range, the pulse width was reduced from 500 to 200 ns in order to minimize the internal heating within a single pulse. The maximum WPE of this device at room temperature is 8.8% at a current density of 2 kA/cm^2. Figure 3.19b illustrates the exponential fit to the threshold current density, $J_{th} = J_0 \exp(T/T_0)$, where a T_0 of 140 K and a J_0 of 0.1 kA/cm^2 within the tested temperature range is obtained. Thus, the maximum operating temperature of the device according to T_0, J_0, and the rollover current density J_{ro} can be estimated to be $T_{max} = T_0 \ln(J_{ro}/J_0)$. For a J_{ro} of 4 kA/cm^2, the maximum operating temperature of 243°C is obtained.

3.5.1 Photonic Crystal DFB QCLs

Broad area (ridge width greater than 50 μm) QCLs are expected to exhibit CW operation at room temperature while these lasers with the conventional FP cavity suffer from a wide emitting spectrum and a broad far field profile, due to the appearance of higher order lateral modes. In order to deal with the spectral and spatial problems of the conventional broad area QCL, photonic crystal DFB (PCDFB) mechanism has introduced [57, 58]. Figure 3.20a shows an SEM image of the bonded laser chip with wire bonds where three PCDFB grating periods have monolithically fabricated on the same wafer. The corresponding lasing spectrum from the three separate emitters is depicted in Fig. 3.20b.

Fig. 3.19 a Temperature dependent power–current–voltage and **b** threshold current density of the 10.2 μm high-power QCL operating in pulsed mode at temperatures between 298 and 373 K [54]

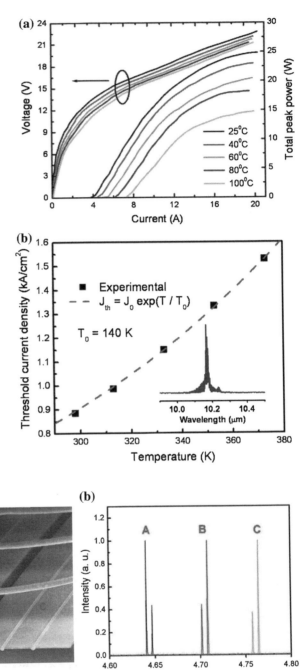

Fig. 3.20 a SEM image of a PCDFB laser chip with three different grating periods and **b** experimentally measured lasing spectrum corresponding to the three grating periods [54]

3.5 High Power QCLs

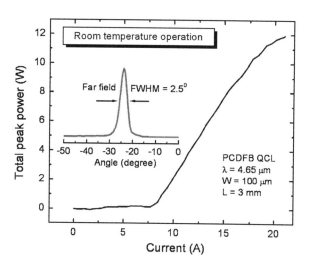

Fig. 3.21 Output power performance of a PCDFB QCL at room temperature in pulsed mode. The inset depicts the far-field characteristics [54]

The output power performance of a PCDFB QCL at 4.65 μm and at room temperature is shown in Fig. 3.21 where the laser is driven in pulsed mode with 1% duty cycle. As it is obvious, a peak power of about 12 W is obtained. The inset also depicts the far field characteristics. The diffraction limited FWHM is 2.5°, which is not current dependent above threshold.

3.6 Dual-Wavelength Generation Based on Monolithic THz-IR QCL

3.6.1 Linear Frequency-Doubling in IR QCL

Parallel to development of QCLs, structures with multi-peaked gain spectra and associated multiple-wavelength emission (multicolor operation) have already been achieved by direct generation [59, 60] or by nonlinear optics [61, 62] in the mid-IR spectral region ($\hbar\omega_{LO\text{-Phonon}} < h\nu$). For applications such as differential spectroscopy, wavelength division multiplexing in free space, like dual wavelength imaging [63], intracavity frequency mixing [64], trace gas sensing where the second wavelength could be used as in-built reference channel [65] and many more complex sensing applications, such as sensing of vapors of large and organic molecules or liquids [66], a light source with emission at multi wavelengths is needed. On the other hand, the absence of re-absorption for widely spaced photon energies make inter-subband devices ideal systems for multi frequency operation [67].

The second-harmonic generation (SHG) in QCLs can be a favorable access way to multi-wavelength operation in mid-infrared region. SHG in QCLs has been demonstrated, and is due to the substantial second-order nonlinear susceptibility that can be designed into QCLs brought about by the large dipole matrix elements associated with intersubband transitions [68, 69]. However, SHG-based two-color

QCLs suffer from the high power difference between the first and the second harmonics generated in the laser cavity (e.g. a linear output power of about 0.5 W and SHG power of about 300 µW as reported in [68]). This high difference between linear and nonlinear power results in poor efficiency of the second harmonic power in the related applications. So, designing dual-color QCL structures with close output powers will be interesting.

Figure 3.22 shows one and half period of a QC laser which is capable of emitting two simultaneous wavelengths (13.77 and 6.88 µm) in mid-IR region which can be implemented by AlGaAs material compositions. The active region of the structure is designed such that the frequency of the first emitted light from a shallow well in a coupled-quantum-well active region is doubled in the adjacent deep well. The Electrons are injected to the active region via a miniband created by SL-like injector and transit to lower energy level by LO-phonon scattering process after emitting the first wavelength. Then, electrons tunnel to the deep well where the second wavelength (6.88 µm) is emitted which is followed by tunneling out to the next injector through an exit barrier and becoming ready for the next active region.

Since the level 4 inside the active region is close to the conduction band edge of the coupling barrier (barrier between two coupled wells), an AlAs blocking barrier [70] is introduced to the coupling barrier to enhance the effective barrier height. A proper design of active region will allow the wavefunctions of the electrons to localize in shallow and deep well thus increase the dipole transition matrix element due to increase of the overlapping factor. Since gain coefficient increases with the square of the optical dipole matrix element, therefore the performance of the laser will enhance. Similar design techniques such as double-spike barrier (instead of

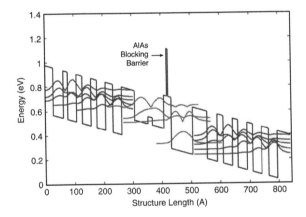

Fig. 3.22 Conduction band profile of the dual-wavelength QCL at an electric field of 70 kV/cm including wavefunctions in the injector minibands and two-coupled quantum well active region. The layer sequences of one active region and injector from left to right are: 50, 35, 20, 30, 20, 30, 20, 30, 25, 30, 46, 45, 15, 40, 10, 5, 10, 72, in angstrom. Layers in *bold* are well regions and layers in *italic* are Si doped layers to $n = 1 \times 10^{17}$ cm^{-3} [90]

3.6 Dual-Wavelength Generation Based on Monolithic THz-IR QCL

single-spike blocking barrier) as introduced for quantum well photo detectors [71] may also be implemented to enhance the carrier confinement in QCLs.

It should be mentioned that for an electron to escape from a quantum well, two major mechanisms are considered: thermionic emission and tunneling. The lifespans due to thermionic emission and tunneling can be expressed as:

$$(\tau_E)_i = \left(\frac{2\pi m_i L_w^2}{K_B T}\right) \exp\left(\frac{V_i(F)}{K_B T}\right) \tag{3.41}$$

$$(\tau_T)_i = \left(\frac{2 m_i L_w^2}{\pi \hbar}\right) \exp\left(\frac{2 L_{ex} \sqrt{2 m_{bi} V_i'F}}{\hbar}\right), \tag{3.42}$$

where K_B, m_i, $V_i(F)$, L_{ex} and $V_i'(F)$ are the Boltzmann constant, the electron effective mass in the subband i, the potential height for electrons in subband i to leap over, the thickness of the exit barrier and the potential for electrons in the subband i to tunnel. In the presence of the blocking barrier, the tunneling and emission probabilities out of level 4 to the next well or continuum is reduced highly due to high effective potential of the blocking barrier (The operation temperature of the structure limited by thermionic emission is about 150 K).

Simulations shows that the continuum states in the applied electric field are not in resonance with injector states hence, they do not interfere directly in QCL transitions and performance. Proper design of the active region localizes highly both of lasing states (i.e. states |4⟩ and |3⟩ and, states |2⟩ and |1⟩) in the corresponding well which results in higher dipole transition matrix element.

To calculate the carrier lifetimes in each of the lasing states it should be mentioned that electron relaxation dynamics is governed predominantly by the electrons interaction with polar optical phonons. The electron–phonon density rates are calculated using Fermi's golden rule as a transition from the initial subband |i⟩ with energy E_i to the final subband |j⟩ with energy E_j. The formalism used for LO-phonon scattering is based on that of Smet et al. [72]. The total scattering rate for absorption of LO-phonons for initial state $\left|i, \vec{k}_i\right\rangle$ based on WKB approximation can be written as:

$$W_{i\to f}^{em}\left(\vec{k}_i\right) = \frac{m^* e^2 \omega_{LO}}{2\hbar^2}\left(\frac{1}{\varepsilon_\infty} - \frac{1}{\varepsilon_0}\right)(n_{\omega LO} + 1)\int_0^{2\pi} B_{i\to f}(q_\perp) d\theta, \tag{3.43}$$

where $n_{\omega LO}$ is the Bose–Einstein distribution for the phonons, ε_s and ε_∞ are the static and high frequency dielectric constants, and $\hbar\omega_{LO}$ is the LO-phonon energy. The expression for $B_{i\to f}$ is:

$$B_{i\to f}(q_\perp) = \int_{-\infty}^{\infty} dx \int_{-\infty}^{\infty} dx' \psi_f^*(x)\psi_i(x)\psi_i^*(x')\psi_f(x') I(x, x', q_\perp), \tag{3.44}$$

where the envelope functions $\psi_{i,f}(x)$ for the initial or final subband can be obtained through solving Schrödinger's equation and $I(x, x', q_\perp)$ is expressed by:

$$I(x, x', q_\perp) = \left[\frac{1}{(q_\perp^2 + q_S^2)^{1/2}} - \frac{|z - z'|q_S^2/2}{(q_\perp^2 + q_S^2)} - \frac{q_S^2/2}{(q_\perp^2 + q_S^2)^{3/2}} \right] \times \exp\left[-\sqrt{q_\perp^2 + q_S^2}|x - x'| \right], \tag{3.45}$$

where $q_\perp$ is the in-plane momentum, and q_S is inverse screening length, which accounts for the screening of electron–electron interactions involving in the electron–LO phonon scattering process. The screening becomes significant when the electron density increases above 10^{11} cm^{-2}. In Eq. 3.43, θ is the angle between the initial and final states characterized by wave-vectors $\vec{k}_i$ and $\vec{k}_f$, and thus the θ—integration corresponds to the integration over all possible final states. The exchanged in-plane momentum $q_\perp$ can be expressed using the momentum conservation equation expressed by:

$$q_\perp^2 = \left|\vec{k}_i - \vec{k}_f\right|^2 = k_i^2 + k_f^2 - 2k_ik_f \cos\theta. \tag{3.46}$$

The magnitude of the in-plane momentum of final states can be derived from energy conservation as below:

$$k_f^2 = k_i^2 + \frac{2m^*\left(E_f - E_i + \hbar\omega_{LO}\right)}{\hbar^2}. \tag{3.47}$$

The total scattering time between subbands $\tau_{i \to f}$ can then be obtained by averaging over all possible initial states in the subband:

$$\frac{1}{\tau_{i \to f}} = \frac{\int_0^\infty dE_k \, \rho_C^{2D}(E_k) f_i(E_k) W_{i \to f}^{em}(E_k)}{\int_0^\infty dE_k \, \rho_C^{2D}(E_k) f_i(E_k)}. \tag{3.48}$$

The quasi-Fermi distribution of the initial state contributes significantly to the average scattering time. It will change with the population at the subband. In the initial calculation, it is assumed that the population in the lower subband is zero. The obtained scattering times are: $\tau_{4-3} = 2$ ps, $\tau_{3-2} = 0.1$ ps and $\tau_{2-1} = 2.8$ ps.

The first requirement for the gain calculation is the dipole transition matrix element calculation in the structure. The obtained values for the matrix elements between electron states are: $\langle z_{12} \rangle = 1.1$ nm, $\langle z_{23} \rangle = 1.97$ nm, $\langle z_{34} \rangle = 2.68$ nm and $\langle z_{13} \rangle = 0.07$ nm. The active region gain parameter in a QCL can be explained by multiplying a gain coefficient (g_0) to the difference of carrier sheet densities of the lasing levels such as [73]:

$$g_0^{ij} = \frac{e^2 E_{ij} |z_{ij}|^2}{2\hbar c \varepsilon_0 n_{\text{eff}} L_p} \frac{\gamma_{ij}}{(E_j - E_i - \hbar\omega)^2 + \gamma_{ij}^2/4} \tag{3.49}$$

3.6 Dual-Wavelength Generation Based on Monolithic THz-IR QCL

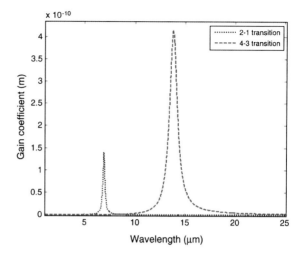

Fig. 3.23 Gain coefficients for $|2\rangle \to |1\rangle$ (6.88 μm) and $|4\rangle \to |3\rangle$ (13.77 μm) transitions [90]

and

$$g(\hbar\omega, N_j - N_i) = g_0^{i,j}(N_j - N_i), \quad (3.50)$$

where i and j are initial and final states respectively. The obtained results for the gain coefficients of both wavelengths are presented in Fig. 3.23.

Since a complete analysis of a QCL requires consideration of both active and injector regions, it involves a large number of rate equations which are cumbersome to deal with. In order to reduce the complexity, a four-level model in two-coupled-well active region with lifetimes of the carriers which takes into account the removal of the carriers is applied here. The cross-gain effects (the influence of λ_1 gain on λ_2 and vice versa) have also considered in the rate equations which become much important whenever the two lasing wavelengths are close to each other. The semiclassical rate equation model of the active region is described in Eqs. 3.51–3.56.

The terms $g_{\lambda_1}^{w_1}$ and $g_{\lambda_2}^{w_2}$ correspond to the gain coefficients of λ_1 (13.77 μm) in the shallow well and λ_2 (6.88 μm) in the deep well respectively. The inserted cross gain-coefficient effects are presented with $g_{\lambda_1}^{w_2}$ and $g_{\lambda_2}^{w_1}$ terms which stand for the gain coefficient of each well in the emission wavelength of the other.

$$\frac{dN_4}{dt} = \eta_{in}\frac{J_{in}}{q} - \frac{N_4}{\tau_{43}} - \frac{N_4}{\tau_{42}} - \frac{\Gamma}{N_p}g_{\lambda_1}^{w_1}\frac{c}{n_{eff}}(N_4 - N_3)P_{\lambda_1}$$
$$- \frac{\Gamma}{N_p}g_{\lambda_2}^{w_1}\frac{c}{n_{eff}}(N_4 - N_3)P_{\lambda_2} \quad (3.51)$$

$$\frac{dN_3}{dt} = \frac{N_4}{\tau_{43}} - \frac{N_3}{\tau_{32}} + \frac{\Gamma}{N_p}g_{\lambda_1}^{w_1}\frac{c}{n_{eff}}(N_4 - N_3)P_{\lambda_1}$$
$$+ \frac{\Gamma}{N_p}g_{\lambda_2}^{w_1}\frac{c}{n_{eff}}(N_4 - N_3)P_{\lambda_2} \quad (3.52)$$

$$\frac{dN_2}{dt} = \frac{N_4}{\tau_{42}} + \frac{N_3}{\tau_{32}} - \frac{N_2}{\tau_{21}} - \frac{\Gamma}{N_p} g^{w_2}_{\lambda_1} \frac{c}{n_{\text{eff}}}(N_2 - N_1)P_{\lambda_1}$$
$$- \frac{\Gamma}{N_p} g^{w_2}_{\lambda_2} \frac{c}{n_{\text{eff}}}(N_2 - N_1)P_{\lambda_2} \quad (3.53)$$

$$\frac{dN_1}{dt} = \frac{N_2}{\tau_{21}} + \frac{N_3}{\tau_{31}} + \frac{\Gamma}{N_p} g^{w_2}_{\lambda_1} \frac{c}{n_{\text{eff}}}(N_2 - N_1)P_{\lambda_1}$$
$$+ \frac{\Gamma}{N_p} g^{w_2}_{\lambda_2} \frac{c}{n_{\text{eff}}}(N_2 - N_1)P_{\lambda_2} - \frac{N_1}{\tau_{\text{tunneling}}} \quad (3.54)$$

$$\frac{dP_{\lambda_1}}{dt} = \Gamma g^{w_1}_{\lambda_1} \frac{c}{n_{\text{eff}}}(N_4 - N_3)P_{\lambda_1} + \Gamma g^{w_2}_{\lambda_1} \frac{c}{n_{\text{eff}}}(N_2 - N_1)P_{\lambda 1} - \frac{P_{\lambda_1}}{\tau_{P_{\lambda_1}}} \quad (3.55)$$

$$\frac{dP_{\lambda_2}}{dt} = \Gamma g^{w_1}_{\lambda_2} \frac{c}{n_{\text{eff}}}(N_4 - N_3)P_{\lambda_2} + \Gamma g^{w_2}_{\lambda_2} \frac{c}{n_{\text{eff}}}(N_2 - N_1)P_{\lambda_2} - \frac{P_{\lambda_2}}{\tau_{P_{\lambda_2}}}. \quad (3.56)$$

In the above equations, N_i's are the carrier sheet densities, η_{in}, J_{in}, Γ, τ_p are the injection efficiency of input current into active region (state $|4\rangle$), input current density in kA/cm^2, confinement factor, and photon life time respectively, τ_{ij}'s are carrier scattering times and P_{λ_1}, P_{λ_2} are the photon sheet densities of the two emitted wavelengths. N_p is number of the periods which are equal to 25. Equations 3.51–3.54 specify the population of a specific subband in terms of the populations of all the other subbands, the scattering rates between the states and also photon generation rates in two different wavelengths.

The first term in Eq. 3.51 describes the current injection rate from injector to active region which populate the state $|4\rangle$. The second and third terms describe the scattering rates between the state $|4\rangle$ and two other close states (state $|3\rangle$ and $|2\rangle$) in the active region where τ_{43} and τ_{42} represent the average time for an electron in state $|4\rangle$ to scatter to state $|3\rangle$ and $|2\rangle$ respectively. The two last terms indicate the photon generation rates in the first well of active region in two different wavelengths. Equations 3.55 and 3.56 show the photon generation rates. The effect of external electric field has been considered in the self-consistent Schrödinger–Poisson analyze of the structure which leads to modifying of energy levels, wavefunctions and consequently scattering times and gain spectra. The gain coefficients can be expressed as:

$$g_{\lambda_1} = \frac{2e^2 E_{43}|z_{43}|^2}{\hbar c \varepsilon_0 n_{\text{eff}} L_p \gamma_{43}}, \quad g_{\lambda_2} = \frac{2e^2 E_{21}|z_{21}|^2}{\hbar c \varepsilon_0 n_{\text{eff}} L_p \gamma_{21}}, \quad (3.57)$$

where E_{ji} is the energy difference between levels i and j, z_{ji} is the dipole matrix element, L_p is one period length and γ_{ji} is the broadening factor. The photon lifetime τ_p in the laser cavity is proportional to the intrinsic loss ($\alpha_i = 12$ cm^{-1}) and mirror loss ($\alpha_m = 4.4$ cm^{-1}) inside the cavity defined by:

3.6 Dual-Wavelength Generation Based on Monolithic THz-IR QCL

$$\tau_p = \frac{n_{\text{eff}}}{c} \frac{1}{\alpha_i + \alpha_m}. \quad (3.58)$$

Therefore, the threshold current can be written as:

$$\Gamma g_{\max}(J_{\text{th}}) = \alpha_i + \alpha_m, \quad J_{\text{th}} = \frac{\alpha_i + \alpha_m}{\Gamma g_0}. \quad (3.59)$$

Finally, the laser output power for each wavelength can be calculated as described below for λ_1 (a similar relation can be obtained for λ_2):

$$P(I) = \eta_{\text{opt}} \frac{E_{43}}{e} \frac{\alpha_m}{\alpha_i + \alpha_m} \frac{\tau_{42}}{\tau_{42} + \tau_{43}} \left[\eta_4 \left(1 - \frac{\tau_{32}}{\tau_{43}} \right) \right] P_{\lambda_1}(I - I_{\text{th}}), \quad (3.60)$$

where $I = wLJ$, $I_{\text{th}} = wLJ_{\text{th}}$ ($w = 18$ μm and $L = 1000$ μm are cavity width and length) and η_{opt} is optical efficiency related to the cavity characteristics and is described by:

$$\eta_{\text{opt}} = \frac{(1 - R_1)\sqrt{R_2}}{\left(\sqrt{R_1} + \sqrt{R_2}\right)\left(1 - \sqrt{R_1 R_2}\right)}. \quad (3.61)$$

It has supposed in the calculation that $\eta_4 = \eta_{\text{in}} = 1$ and $\eta_{\text{opt}} = 1/2$. Figure 3.24 shows the simulated results for output power of QCL where the powers for λ_1 and λ_2 are illustrated with dash and straight lines respectively. As it is obvious, the optical output powers of the two wavelengths are in same orders. The noticeable point is that the output power of λ_2 wavelength is approximately twice as λ_1 in the 1 A bias current. This phenomenon contrasts with the idea of using the SHG effect where the second harmonic power is always much weaker than fist harmonic power. Also, the observed high slope of the power–current curve near the threshold current is almost in accordance with reported experimental results [73].

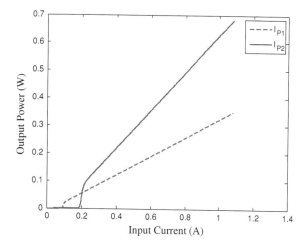

Fig. 3.24 Output powers of two emitted wavelengths in QCL as a function of input bias current (I_{P1} stands for the power of λ_1 and I_{P2} is for λ_2) [90]

3.6.2 Dual Color Terahertz QCL

Terahertz QC lasers based on GaAs/AlGaAs quantum wells have, in fact, been developed in the past few years and currently span the 1.2–5 THz range with peak output powers often on the order of a few tens milliwats [74–77] and maximum temperatures near 170 and 120 K for pulsed and continuous wave emission, respectively [19]. The operation is limited due to the presence of LO-phonons with energies $\hbar\omega_{LO}$ ($\sim$36 meV) relatively close to the room temperature thermal energy $K_B T$ of 26 meV. Hence at or near room temperature the process of thermally activated LO-phonon scattering and backfilling makes a dramatic reduction in the device population inversion and optical gain [78]. Obviously, a material system with large LO-phonon energy will be desirable for the high-temperature operation of THz QCLs. One promising system which has attracted considerable attention in recent years is the GaN/AlGaN quantum wells, where the LO-phonon energy is $\sim$91.9 meV [79]. Previous theoretical studies have suggested that even room-temperature terahertz lasing maybe feasible with III-nitride QWs [80, 81]. Moreover, QCLs based on GaAs/AlGaAs are not capable of emitting in the energy range around LO-phonon energies, leaving the gap in the spectral scale between 30 and 40 μm [82]. This limitation can also overcome by significantly larger LO-phonon energy of GaN/AlGaN system.

Parallel to development of QCLs, structures with multi-peaked gain spectra and associated multiple-wavelength emission (multicolor operation) have studied in the mid-IR spectral region. However, these developments are not seen in the terahertz frequencies except some structures with dramatically low operation temperature or needy to apply huge magnetic fields [83]. This is because in terahertz regime, energy subbands are so close together and hence their selectively population and depopulation becomes very difficult.

Figure 3.25 illustrates two and half periods of active region and relevant wave functions of a terahertz QCL based on GaN/Al$_{0.46}$Ga$_{0.54}$N material system, with resonant-phonon-assisted depopulation scheme which is capable of emitting at two widely separated wavelengths (33 and 52 μm), located in terahertz frequency range. The structure contains 50 repetition of GaN multiple quantum well periods. Each period consists of four wells for gain achievement followed by another four wells for injection process. An electric field of 116 kV/cm is applied to have the first radiative transition $4 \rightarrow 3$, at $\lambda_1 = 33$ μm ($\Delta E_{43} = 38$ meV, $v_1 = 9.1$ THz) and second one $2 \rightarrow 1$, at $\lambda_2 = 52$ μm ($\Delta E_{21} = 24$ meV, $v_2 = 5.8$ THz). Then, electrons flow through an injector mini-band and exactly inject into level 4 of the next period. The principal challenge in designing terahertz QCLs is difficulty of selectively populate upper lasing level and selectively depopulate lower laser level which related to very low energy difference between radiative sub-bands. To abate this problem in the structure presented in Fig. 3.25, wave functions have weak overlapping and all transitions are designed to be diagonal. This feature also helps to suppress electrons thermal backfilling and enhance temperature characteristics. By this feature, the lower (upper) level is

3.6 Dual-Wavelength Generation Based on Monolithic THz-IR QCL

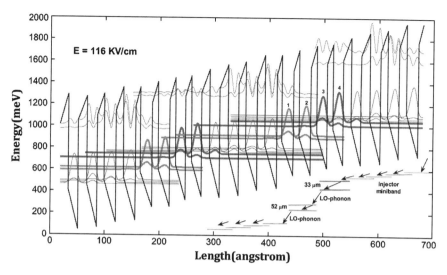

Fig. 3.25 Conduction band diagram and moduli squared of the electron wavefunctions of the GaN/Al$_{0.46}$Ga$_{0.54}$N structure for two gain sections embedded between three injector minibands. 4 → 3 and 2 → 1 are radiative transitions at $\lambda_1 = 33$ μm and $\lambda_2 = 52$ μm respectively. *Inset* depicts the simplified illustration of optical transitions, carrier injection and extraction procedures. The layer sequence of one period of the structure from right to left, starting from the first well of active region corresponding to wavefunction 4, in angstroms is: 18, 11, 18, 11, 21, 11, 21, 9, 22, 14, 21, 13, 19, 16, 18 and 15 [91]

depopulated (populated) without depopulating (populating) the upper (lower) level as well.

To satisfy the population inversion condition between two involved radiative sub-bands for gain attainment, extraction of carriers from lower level must be more rapid than re-population by carriers, which have been scattered from upper level. For this purpose, energy spaces equivalent to LO-phonon energy (~ 91.9 meV) is embedded among lower radiative levels i.e. levels 3, 1 and the next energy sub-bands to extract electrons via ultra-fast LO-phonon scattering mechanism.

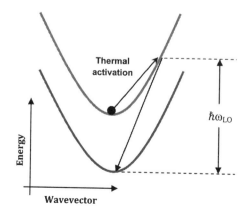

Fig. 3.26 Illustration of thermally activated phonon scattering mechanism which can be suppressed by a large LO-phonon energy [91]

Moreover, this energy space acts like a barrier in front of thermally activated phonon scattering (Fig. 3.26) and backfilling [84, 85] which plays an important role in temperature properties improvement in GaN quantum wells.

When a confined level gets closer to the top of the barriers, the probability of thermal activation of electrons to continuum states increases. Another interest in this structure is very high barriers and large energy space between involved lasing states and those close to the continuum (about 350 meV) which make it almost impossible for electrons to escape or thermally activate to upper states even in high temperatures and hence suppress relative leakage current and improve temperature specifications and injection efficiencies. With aforesaid temperature characteristics, the dominant factor that can be a criterion for maximum operating temperature of the active region is the temperature in which, electrons have enough thermal energy $K_B T \sim \Delta E_{21} = 24$ meV to backfill level 2 and hamper the system operation. This temperature is about 278 K.

Solving the simplified rate equations for the five-level model of the energy levels of the QC laser structure, schematically illustrated in Fig. 3.27, will lead to obtaining the system dynamics. Here, states 4 and 3 are upper and lower radiative levels of the first optical transition at $\lambda_1 = 33$ μm, into which electrons are injected with the rate of η_1 and $(1 - \eta_1)$ respectively.

Levels 2 and 1 are related to the second optical transition at $\lambda_2 = 52$ μm. It is supposed that the total outflow rates of levels 4 and 3 ($1/\tau_{42} + 1/\tau_{41} + 1/\tau_{32} + 1/\tau_{31}$) is equal to the inject current into levels 2 and 1 with respective rates of η_2 and $(1 - \eta_2)$. η is the injection efficiency and i is the reservoir (injector) state.

The rate equations can be determined as function of various radiative and non-radiative scattering lifetimes as:

$$\frac{dn_4}{dt} = \frac{\eta_1 I N_P}{|e|} - \frac{n_4}{\tau_{41}} - \frac{n_4}{\tau_{43}} - \frac{n_4}{\tau_{42}} - \frac{n_4}{\tau_{\mathrm{spl}}} \frac{\Gamma_1}{N_P} - (n_4 - n_3) \frac{\Gamma_1}{N_P} \frac{n_{\mathrm{ph}}(\hbar\omega_1)}{\tau_{\mathrm{sp}}(\hbar\omega_1)} \quad (3.62)$$

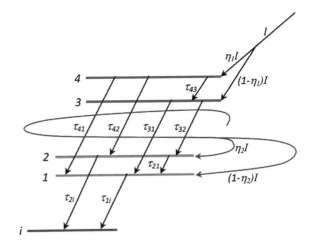

Fig. 3.27 A simple schematic of level diagram and various transition lifetimes to describe the rate equation analysis of a five level QCL. 4, 3, 2 and 1 denote the radiative levels, i is the injector state, I is the injected current, η_1 and η_2 are injection efficiencies and τ_{43}, τ_{42}, τ_{41}, τ_{32}, τ_{31}, τ_{21}, τ_{2i} and τ_{1i} are the lifetimes due to non-radiative transitions $4 \to 3$, $4 \to 2$, $4 \to 1$, $3 \to 2$, $3 \to 1$, $2 \to 1$, $2 \to i$ and $1 \to i$, respectively [91]

3.6 Dual-Wavelength Generation Based on Monolithic THz-IR QCL

$$\frac{dn_3}{dt} = \frac{(1-\eta_1)IN_P}{|e|} - \frac{n_3}{\tau_{32}} - \frac{n_3}{\tau_{31}} + \frac{n_4}{\tau_{43}} + \frac{n_4}{\tau_{sp1}}\frac{\Gamma_1}{N_P} + (n_4 - n_3)\frac{\Gamma_1}{N_P}\frac{n_{ph}(\hbar\omega_1)}{\tau_{sp}(\hbar\omega_1)} \quad (3.63)$$

$$\frac{dn_2}{dt} = \frac{\eta_2 IN_P}{|e|} - \frac{n_2}{\tau_{21}} - \frac{n_2}{\tau_{2i}} - \frac{n_2}{\tau_{sp2}}\frac{\Gamma_2}{N_P} - (n_2 - n_1)\frac{\Gamma_2}{N_P}\frac{n_{ph}(\hbar\omega_2)}{\tau_{sp}(\hbar\omega_2)} \quad (3.64)$$

$$\frac{dn_1}{dt} = \frac{(1-\eta_2)IN_P}{|e|} - \frac{n_1}{\tau_{1i}} + \frac{n_4}{\tau_{41}} + \frac{n_2}{\tau_{21}} + \frac{n_2}{\tau_{sp2}}\frac{\Gamma_2}{N_P} + (n_2 - n_1)\frac{\Gamma_2}{N_P}\frac{n_{ph}(\hbar\omega_2)}{\tau_{sp}(\hbar\omega_2)} \quad (3.65)$$

$$\frac{dn_{ph}(\hbar\omega_1)}{dt} = (n_4 - n_3)\frac{\Gamma_1}{N_P}\frac{n_{ph}(\hbar\omega_1)}{\tau_{sp}(\hbar\omega_1)} - \frac{n_{ph}(\hbar\omega_1)}{\tau_{ph1}} \quad (3.66)$$

$$\frac{dn_{ph}(\hbar\omega_2)}{dt} = (n_2 - n_1)\frac{\Gamma_2}{N_P}\frac{n_{ph}(\hbar\omega_2)}{\tau_{sp}(\hbar\omega_2)} - \frac{n_{ph}(\hbar\omega_2)}{\tau_{ph2}}. \quad (3.67)$$

The notation used in this derivation is summarized below:

- I is the electronic current (in the direction of electron transport) flowing through the device, N_P is the number of repeated modules in the QCL and $\Gamma_{1,2}$ is the fraction of the optical mode that overlaps with the entire active region or confinement factor. Up to the end, subscripts 1 and 2 will represent the parameter in the first ($\lambda_1 = 33$ μm) and second ($\lambda_2 = 52$ μm) modes, respectively.
- n_4, n_3, n_2 and n_1 are the electron populations in the subbands 4, 3, 2 and 1 respectively, summed over all QCL modules and $n_{ph}(\hbar\omega_{1,2})$ is the number of photons per unit energy (1/J) generated into lasing modes, due to stimulated optical transitions.
- $\tau_{sp1,2}$ is the spontaneous emission lifetime into all available cavity modes, $\tau_{sp}(\hbar\omega_{1,2})$ is the spontaneous emission lifetime into the lasing modes $\omega_{1,2}$ and the last terms in Eqs. 3.62–3.65 are the stimulated emission rates which are related to the number of inverted electrons stimulated by photons with energy $(\hbar\omega_{1,2})$.
- $\tau_{ph1,2}$ is the photon lifetime in the cavity, which can be expressed as:

$$\frac{1}{\tau_{ph1,2}} = \frac{c(\alpha_{w1,2} + \alpha_m)}{n_{eff1,2}}, \quad (3.68)$$

where $\alpha_{w1,2}$ and α_m are the waveguide and mirror losses (1/m), respectively, c is the speed of light in vacuum, and $n_{eff1,2}$ is the modal effective refractive index.

Below or near lasing threshold, $n_{ph}(\hbar\omega_{1,2})$ is not a considerable value and because $\tau_{sp1,2}$, $\tau_{sp}(\hbar\omega_{1,2})$ are several orders of magnitude greater than non-radiative lifetimes, two last terms in Eqs. 3.62–3.65 can be neglected. In threshold condition, the steady-state subband populations can be obtained as:

$$n_{4th} = I_{th1} \frac{\eta_{th1} N_P}{|e|} \frac{\tau_{42}\tau_{43}}{(\tau_{42}+\tau_{43})} \tag{3.69}$$

$$n_{3th} = I_{th1} \frac{N_P}{|e|} \frac{\tau_3(\tau_{42}+(1-\eta_{th1})\tau_{43})}{(\tau_{42}+\tau_{43})} \tag{3.70}$$

$$n_{2th} = I_{th2} \frac{\eta_{th2} N_P}{|e|} \frac{\tau_{2i}\tau_{21}}{(\tau_{2i}+\tau_{21})} \tag{3.71}$$

$$n_{1th} = I_{th2} \frac{N_P}{|e|} \frac{\tau_{1i}(\tau_{2i}+(1-\eta_{th2})\tau_{21})}{(\tau_{2i}+\tau_{21})}, \tag{3.72}$$

where τ_3 is lifetime of level 3 which defines as $1/\tau_3 = 1/\tau_{32} + 1/\tau_{31}$. It should be noted that the term n_4/τ_{41} in Eqs. 3.62 and 3.65 vanishes because of so low overlapping between states 4 and 1 and consequently so large $4 \to 1$ transition lifetime.

Above threshold, the photon number in lasing modes n_{ph} ($\hbar\omega_{1,2}$) grows up significantly, so, stimulated transition lifetimes in Eqs. 3.62–3.65 can not be neglected in comparison to non-radiative lifetimes. In steady state, the following results are obtained by solving the Eqs. 3.62–3.67:

$$n_4 = n_{4th} + (I - I_{th1}) \frac{N_P}{|e|} \frac{\tau_3\tau_{42}}{(\tau_3+\tau_{42})} \tag{3.73}$$

$$n_3 = n_{3th} + (I - I_{th1}) \frac{N_P}{|e|} \frac{\tau_3\tau_{42}}{(\tau_3+\tau_{42})} \tag{3.74}$$

$$n_2 = n_{2th} + (I - I_{th2}) \frac{N_P}{e} \frac{\tau_{1i}\tau_{2i}}{(\tau_{1i}+\tau_{2i})} \tag{3.75}$$

$$n_1 = n_{1th} + (I - I_{th2}) \frac{N_P}{e} \frac{\tau_{1i}\tau_{2i}}{(\tau_{1i}+\tau_{2i})} \tag{3.76}$$

$$n_{ph}(\hbar\omega_1) = \frac{N_P}{|e|} \tau_{ph1} \left[\frac{(\eta_{th1}\tau_{42}\tau_{43} - \tau_3\tau_{42} - (1-\eta_{th1})\tau_3\tau_{43})}{(\tau_3+\tau_{42})\tau_{43}} \right.$$
$$\left. \times (I - I_{th1}) + I(\eta - \eta_{th1}) \right] \tag{3.77}$$

$$n_{ph}(\hbar\omega_2) = \frac{N_P}{|e|} \tau_{ph2} \left[\frac{(\eta_{th2}\tau_{2i}\tau_{21} - \tau_{1i}\tau_{2i} - (1-\eta_{th2})\tau_{1i}\tau_{21})}{(\tau_{1i}+\tau_{2i})\tau_{21}} \right.$$
$$\left. \times (I - I_{th2}) + I(\eta - \eta_{th2}) \right]. \tag{3.78}$$

3.6 Dual-Wavelength Generation Based on Monolithic THz-IR QCL

To obtain above populations, at first, scattering times calculation is necessary which is done between two subband i and f following Ferreira and Bastard's approach [85]. Obtained lifetimes are: $\tau_{i\bullet 4} = 0.3$ ps, $\tau_{i\bullet 3} = 1.96$ ps, $\tau_{43} = 0.24$ ps, $\tau_{42} = 41$ ps, $\tau_{32} = 0.14$ ps, $\tau_{31} = 12$ ps, $\tau_{21} = 0.23$ ps, $\tau_{2i} = 7$ ps and $\tau_{1i} = 0.96$ ps where i' is the injector state of previous period which injects electrons into level 4, and $\tau_{i\bullet 4}$, $\tau_{i\bullet 3}$ are corresponding lifetimes due to $i' \rightarrow 4$ and $i' \rightarrow 3$ transitions, respectively.

Large quantities of $\tau_{i\bullet 3}$, τ_{31}, τ_{42} and τ_{2i} which are due to small overlapping between relevant electron wavefunctions, make two prominent features in the designed laser structure which are high injection efficiencies and excellent selectively extractions. In general, it is difficult to determine η computationally in an accurate manner. In the ideal case $\eta = 1$ and deviations from this value can be due either to thermal activation from the injector to continuum or highly excited states which is cancelled in the structure as it explained heretofore, or to direct injection from the injector to lower states [73]. Hence, the following approximations are use for injection efficiencies, $\eta_1 \approx R_{i'4}/R_{i'4}(R_{i'4} + R_{i'3}) \cdot (R_{i'4} + R_{i'3})$, $\eta_2 \approx (R_{32} + R_{42})/(R_{32} + R_{42})(R_{32} + R_{42} + R_{31}) \cdot (R_{32} + R_{42} + R_{31})$, which give $\eta_1 = 0.91$ and $\eta_2 = 0.98$. The label R represents the transition rate (inverse of transition time, τ^{-1}) between involved subbands in subscripts, therefore, numerators are the net injected rates into upper radiative levels and denominators are the whole rate coming from the previous stage.

If a population inversion is established in the system, each optical transition leads to a net stimulated emission into the cavity mode and correspondingly, power would be added to the field as it propagates through the system. The propagation gain (also referred to as material gain) is defined as variation of the photon flux divided by the number of photons per unit length of the structure which can be expressed in the form of following expression (in 1/m) due to radiative transition between upper and lower subbands u and l:

$$g(\omega) = \frac{\pi e^2 \omega}{\varepsilon_0 n_{\text{eff}} c} |z_{ul}|^2 \Delta N_{3D} \frac{\Delta E/(2\pi)}{(E_u - E_l + \hbar\omega)^2 + (\Delta E/2\pi)^2} \quad (3.79)$$

and the peak gain is obtained at the resonance condition $\omega = \omega_{ul} = (E_u - E_l)/\hbar$

$$g(\omega_{ul}) = \frac{2e^2}{\hbar\varepsilon_0 n_{\text{eff}} c} |z_{ul}|^2 \Delta N_{3D} \frac{v_{ul}}{\Delta v}. \quad (3.80)$$

The last term in Eq. 3.79 represents the broadening in the gain spectrum that is approximated by a normalized Lorentzian function, where ΔE is the FWHM. ΔN_{3D} is the 3D population inversion density computed over volume of all QCL modules, $v_{ul} = \omega_{ul}/2\pi$ and z_{ul} is the dipole matrix element given by $z_{ul} = \int_{-\infty}^{+\infty} \psi_u(z) z \psi_l(z) dz$, where the calculated results are $z_{43} = 0.88$ nm and $z_{21} = 1.09$ nm. In threshold condition, the optical gain becomes equal to losses $g(\omega_{ul}) = \alpha_m + \alpha_w$. Combining Eqs. 3.69–3.72 and 3.79, threshold current densities for both wavelengths are obtained as $J_{th1} = 0.98$ kA/cm^2 and $J_{th2} = 0.58$ kA/cm^2 corresponding to currents $I_{th1} = 197$ mA and $I_{th2} = 116$ mA, respectively.

Here, some of the parameters used in the simulations are listed: $n_{eff1,2} = 2.22$, $N_p = 50$, $\gamma_{43} = 3$ meV, $\gamma_{21} = 3.2$ meV, $\alpha_m = 10$ cm^{-1}, $\alpha_{w1} = 20$ cm^{-1}, $\alpha_{w2} = 25$ cm^{-1}, and the structure dimensions are 50 nm × 6 μm × 2000 μm. n_{eff} is similar for both wavelengths because of flat refractive index spectra of GaN in long wavelengths [86]. The FWHM value has an inverse relation with individual subbands lifetimes, so because of larger lifetimes of levels 4 and 3 in comparison with levels 2 and 1, γ_{43} becomes greater than γ_{21} and α_{w1} is smaller than α_{w2} because of larger carrier absorption losses in longer wavelengths.

The level population densities and relevant population inversions below and above lasing threshold are displayed in Fig. 3.28. As its shown, above the threshold, the populations of the individual subbands increase linearly with current, but $N_4 - N_3$ and $N_2 - N_1$ do not change from its value at threshold. This is because of inefficient extraction of electrons from the lower subbands, which does not allow all extra current being pumped into modules beyond threshold to be converted into photons.

The optical output power is a useful measure explaining the performance of the laser. Due to the cascading scheme, it is intrinsically high in QCLs because the electrons that have contributed in generating photons in one module of the active region are still present in conduction band and can be reused in subsequent modules of the active region. The power performance of the lasers shows the validity of the cascading scheme [87]. The out-coupled powers at two wavelengths can be calculated through:

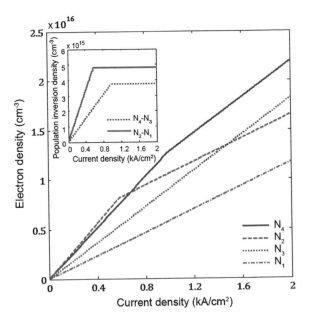

Fig. 3.28 Population densities of the radiative levels and relevant population inversion densities (*inset*) as a function of current per unit cross-sectional area [91]

3.6 Dual-Wavelength Generation Based on Monolithic THz-IR QCL

$$P(\hbar\omega_1) = \hbar\omega_0 n_{ph}(\hbar\omega_1)\alpha_m \frac{c}{n_{eff1}}$$

$$= \frac{N_p \hbar\omega_0}{|e|} \frac{\alpha_m}{\alpha_{w1} + \alpha_m} \left[\frac{(\eta_{th1}\tau_{42}\tau_{43} - \tau_3\tau_{42} - (1 - \eta_{th1})\tau_3\tau_{43})}{(\tau_3 + \tau_{42})\tau_{43}} \right.$$

$$\left. \times (I - I_{th1}) + I(\eta - \eta_{th1}) \right] \tag{3.81}$$

$$P(\hbar\omega_2) = \hbar\omega_0 n_{ph}(\hbar\omega_2)\alpha_m \frac{c}{n_{eff2}}$$

$$= \frac{N_p \hbar\omega_0}{|e|} \frac{\alpha_m}{\alpha_{w2} + \alpha_m} \left[\frac{(\eta_{th2}\tau_{2i}\tau_{21} - \tau_{1i}\tau_{2i} - (1 - \eta_{th2})\tau_{1i}\tau_{21})}{(\tau_{1i} + \tau_{2i})\tau_{21}} \right.$$

$$\left. \times (I - I_{th2}) + I(\eta - \eta_{th2}) \right], \tag{3.82}$$

where η_{th} varies in different bias conditions. Taking this into account, the power–current treatment of the laser along with the gain spectrum of both wavelengths are presented in Fig. 3.29. Absolutely, the output power slakes when the bias gets far away from its optimum value and decrease the slope efficiency d_P/d_I in the P–I curve. The maximum output powers are $P_{1max} = 46$ mW and $P_{2max} = 76$ mW for two wavelengths, respectively.

As it can be seen, the magnitudes of gain spectrums are close together and output powers have almost similar behaviors. Therefore, unlike nonlinear processes this advantage is useful in applications requiring the same level of output powers for both wavelengths.

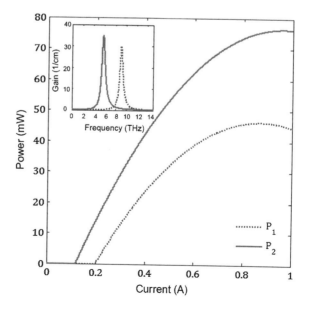

Fig. 3.29 Output power curves versus bias current. The *inset* represents the gain spectrum [91]

References

1. Helm, M., England, P., Colas, E., DeRosa, F., Allen, S.J.: Intersubband emission from semiconductor superlattices excited by sequential resonant tunneling. Phys. Rev. Lett. **63**(1), 74–77 (1989)
2. Köhler, R., Tredicucci, A., Beltram, F., Beere, H.E., Linfield, E.H., Davies, A.G., Ritchie, D.A.: High-intensity interminiband terahertz emission from chirped superlattices. Appl. Phys. Lett. **80**(11), 1867–1869 (2002)
3. Blaser, S., Rochat, M., Beck, M., Faist, J., Oesterle, U.: Far-infrared emission and stark-cyclotron resonances in a quantum cascade structure based on photon-assisted tunneling transition. Phys. Rev. B **61**(12), 8369–8374 (2000)
4. Kazarinov, R.F., Suris, R.A.: Possibility of the amplification of electromagnetic waves in a semiconductor with a superlattice. Sov. Phys. Semicond. **5**(4), 707–709 (1971)
5. Kazarinov, R.F., Suris, R.A.: Electric and electromagnetic properties of semiconductors with a superlattice. Sov. Phys. Semicond. **6**(1), 120–131 (1972)
6. Faist, J., Capasso, F., Sivco, D.L., Sirtori, C., Hutchinson, A.L., Cho, A.Y.: Quantum cascade laser. Science **264**, 553–556 (1994)
7. Köhler, R., Tredicucci, A., Beltram, F., Beere, H.E., Linfield, E.H., Davies, A.G., Ritchie, D.A., Iotti, R.C., Rossi, F.: Terahertz semiconductor-heterostructure laser. Nature **417**, 156–159 (2002)
8. Ulrich, J., Zobl, R., Finger, N., Unterrainer, K., Strasser, G., Gornik, E.: Terahertz-electroluminescence in a quantum cascade structure. Physica B **272**, 216–218 (1999)
9. Sirtori, C., Faist, J., Capasso, F., Sivco, D.L., Hutchinson, A.L., Cho, A.Y.: Quantum cascade laser with plasmon-enhanced waveguide operating at 8.4 μm wavelength. Appl. Phys. Lett. **66**(24), 3242–3244 (1995)
10. Williams, B.S.: Terahertz quantum-cascade lasers. Nat. Photon. (Rev. Artic.) **1**, 517–525 (2007). www.nature.com/naturephotonics
11. Rochat, M., Ajili, L., Willenberg, H., Faist, J., Beere, H., Davies, G., Linfield, E., Ritchie, D.: Low-threshold terahertz quantum-cascade lasers. Appl. Phys. Lett. **81**(8), 1381–1383 (2002)
12. Tredicucci, A., Capasso, F., Gmachl, C., Sivco, D.L., Hutchinson, A.L., Cho, A.Y., Faist, J., Scamarcio, G.: High-power inter-miniband lasing in intrinsic superlattices. Appl. Phys. Lett. **72**(19), 2388–2390 (1998)
13. Helm, M.: Infrared spectroscopy and transport of electrons in semiconductor superlattices. Semicond. Sci. Technol. **10**, 557–575 (1995)
14. Köhler, R., Tredicucci, A., Beltram, F., Beere, H.E., Linfield, E.H., Davies, A.G., Ritchie, D.A.: Low-threshold quantum-cascade lasers at 3.5 THz ($\lambda = 85$ μm). Opt. Lett. **28**(10), 810–812 (2003)
15. Faist, J., Beck, M., Aellen, T., Gini, E.: Quantum cascade lasers based on a bound-to-continuum transition. Appl. Phys. Lett. **78**(2), 147–149 (2001)
16. Scalari, G., Ajili, L., Faist, J., Beere, H., Linfield, E., Ritchie, D., Davies, G.: Far-infrared ($\lambda \sim 87$ μm) bound-to-continuum quantum-cascade lasers operating up to 90 K. Appl. Phys. Lett. **82**(19), 3165–3167 (2003)
17. Ajili, L., Scalari, G., Faist, J., Beere, H., Linfield, E., Ritchie, D., Davies, G.: High power quantum cascade lasers operating at $\lambda \sim 87$ μm and 130 μm. Appl. Phys. Lett. **85**(18), 3986–3988 (2004)
18. Williams, B.S., Callebaut, H., Kumar, S., Hu, Q., Reno, J.L.: 3.4-THz quantum cascade laser based on longitudinal–optical–phonon scattering for depopulation. Appl. Phys. Lett. **82**(7), 1015–1017 (2003)
19. Williams, B.S., Kumar, S., Hu, Q., Reno, J.L.: Operation of terahertz quantum-cascade lasers at 164 K in pulsed mode and at 117 K in continuous-wave mode. Opt. Express **13**, 3331–3339 (2005)

20. Kohler, R., Tredicucci, A., Mauro, C., Beltram, F., Beere, H.E., Linfield, E.H., Davies, A.G., Ritchie, D.A.: Terahertz quantum-cascade lasers based on an interlaced photon–phonon cascade. Appl. Phys. Lett. **84**(8), 1266–1268 (2004)
21. Scalari, G., Hoyler, N., Giovannini, M., Faist, J.: Terahertz bound-to-continuum quantum-cascade lasers based on optical–phonon scattering extraction. Appl. Phys. Lett. **86**, 181101-1–181101-3 (2005)
22. Yariv, A.: Quantum Electronics, 3rd edn. Wiley, New York (1989)
23. Sirtori, C., Kruck, P., Barbieri, S., Page, H., Nagle, J., Beck, M., Faist, J., Oesterle, U.: Low-loss al-free waveguides for unipolar semiconductor lasers. Appl. Phys. Lett. **75**(25), 3911–3913 (1999)
24. Strasser, G., Gianordoli, S., Hvozdara, L., Schrenk, W., Unterrainer, K., Gornik, E.: GaAs/AlGaAs superlattice quantum cascade lasers at $\lambda \approx 13$ µm. Appl. Phys. Lett. **75**(10), 1345–1348 (1999)
25. Becker, C., Sirtori, C., Page, H., Glastre, G., Ortiz, V., Marcadet, X., Stellmacher, M., Nagle, J.: GaAs/AlGaAs quantum cascade lasers based on large direct conduction band discontinuity. Appl. Phys. Lett. **77**(4), 463–465 (2000)
26. Wilson, L.R., Keightley, P.T., Cockburn, J.W., Skolnick, M.S., Clark, J.C., Grey, R., Hill, G.: Controlling the performance of GaAs–AlGaAs quantum-cascade lasers via barrier height modifications. Appl. Phys. Lett. **76**(7), 801–804 (2000)
27. Wilson, L.R., Cockburn, J.W., Steer, M.J., Carder, D.A., Skolnick, M.S., Hopkinson, M., Hill, G.: Decreasing the emission wavelength of GaAs–AlGaAs quantum cascade lasers by the incorporation of ultrathin InGaAs layers. Appl. Phys. Lett. **78**(4), 413–415 (2001)
28. Colombelli, R., Capasso, F., Gmachl, C., Hutchinson, A.L., Sivco, D.L., Tredicucci, A., Wanke, M.C., Sergent, A.M., Cho, A.Y.: Far-infrared surface-plasmon quantum-cascade lasers at 21.5 µm and 24 µm wavelengths. Appl. Phys. Lett. **78**(18), 2620–2622 (2001)
29. Sirtori, C., Gmachl, C., Capasso, F., Faist, J., Sivco, D.L., Hutchinson, A.L., Cho, A.Y.: Long-wavelength ($\lambda \approx 8$–11.5 µm) semiconductor lasers with waveguides based on surface plasmons. Opt. Lett. **23**(17), 1366–1368 (1998)
30. Tredicucci, A., Gmachl, C., Wanke, M.C., Capasso, F., Hutchinson, A.L., Sivco, D.L., Chu, S.G., Cho, A.Y.: Surface plasmon quantum cascade lasers at $\lambda \sim 19$ µm. Appl. Phys. Lett. **77**(15), 2286–2288 (2000)
31. Faist, J., Capasso, F., Sirtori, C., Sivco, D.L., Cho, A.Y.: Quantum cascade lasers. In: Liu, H.C., Capasso, F. (eds.) Intersubband Transitions in Quantum Wells: Physics and Device Applications II, vol. 66, Chapter 1, pp. 1–83, Academic Press, San Diego (2000)
32. Gmachl, C., Capasso, F., Tredicucci, A., Sivco, D.L., Hutchinson, A.L., Cho, A.Y.: Long wavelength ($\lambda \sim 13$ µm) quantum cascade lasers. IEE Elect. Lett. **34**(11), 1103–1104 (1998)
33. Tredicucci, A., Gmachl, C., Capasso, F., Sivco, D.L., Hutchinson, A.L., Cho, A.Y.: Long wavelength superlattice quantum cascade lasers at $\lambda = 17$ µm. Appl. Phys. Lett. **74**(5), 638–640 (1999)
34. Tredicucci, A., Gmachl, C., Capasso, F., Hutchinson, A.L., Sivco, D.L., Cho, A.Y.: Single-mode surface-plasmon laser. Appl. Phys. Lett. **76**(16), 2164–2166 (2000)
35. Rochat, M., Beck, M., Faist, J., Oesterle, U.: Measurement of far-infrared waveguide loss using a multisection single-pass technique. Appl. Phys. Lett. **78**(14), 1967–1969 (2001)
36. Faist, J., Gmachl, C., Capasso, F., Sirtori, C., Sivco, D.L., Baillargeon, J.N., Cho, A.Y.: Distributed feedback quantum cascade lasers. Appl. Phys. Lett. **70**(20), 2670–2672 (1997)
37. Gmachl, C., Faist, J., Baillargeon, J.N., Capasso, F., Sirtori, C., Sivco, D.L., Chu, S.G., Cho, A.Y.: Complex-coupled quantum cascade distributed-feedback laser. IEEE Photon. Technol. Lett. **9**(8), 1090–1092 (1997)
38. Gmachl, C., Capasso, F., Faist, J., Hutchinson, A.L., Tredicucci, A., Sivco, D.L., Baillargeon, J.N., Chu, S.G., Cho, A.Y.: Continuous-wave and high-power pulsed operation of index-coupled distributed quantum cascade laser at $\lambda \sim 8.5$ µm. Appl. Phys. Lett. **72**(12), 1430–1432 (1998)
39. Hofstetter, D., Faist, J., Beck, M., Oesterle, U.: Surface-emitting 10.1 µm quantum-cascade distributed feedback lasers. Appl. Phys. Lett. **75**(24), 3769–3771 (1999)

40. Hofstetter, D., Faist, J., Beck, M., Muller, A., Oesterle, U.: Demonstration of high-performance 10.16 μm quantum cascade distributed feedback lasers fabricated without epitaxial regrowth. Appl. Phys. Lett. **75**(5), 665–667 (1999)
41. Schrenk, W., Finger, N., Gianordoli, S., Hvozdara, L., Strasser, G., Gornik, E.: GaAs/AlGaAs distributed feedback quantum cascade lasers. Appl. Phys. Lett. **76**(3), 253–255 (2000)
42. Gmachl, C., Capasso, F., Tredicucci, A., Sivco, D.L., Baillargeon, J.N., Hutchinson, A.L., Cho, A.Y.: High power, continuous-wave, current-tunable, single-mode quantum-cascade distributed-feedback lasers at $\lambda \sim 5.2$ and $\lambda \sim 7.95$ μm. Opt. Lett. **25**(4), 230–232 (2000)
43. Callebaut, H., Kumar, S., Williams, B.S., Hu, Q., Reno, J.L.: Analysis of transport properties of tetrahertz quantum cascade lasers. Appl. Phys. Lett. **83**(2), 207–209 (2003)
44. Hu, Q.: Terahertz quantum cascade lasers and real-time THz imaging. Lasers and Electro-Optics Society, 2007. LEOS 2007. The Annual Meeting of the IEEE Conferences, 858–859 (2007). doi:10.1109/LEOS.2007.4382679
45. Hu, Q.: Terahertz quantum cascade lasers and video-rate THz imaging. In: Infrared and Millimeter Waves, 2007 and the 2007 15th International Conference on Terahertz Electronics, IRMMW-THz., IEEE Conferences, 24–25 (2007)
46. Callebaut, H., Williams, B., Kumar, S., Hu, Q.: Analysis of transport properties of THz quantum cascade lasers. Chapter 34. Terahertz and Infrared Quantum Cascade Lasers, and Real-time Imaging, Sponsors, National Science Foundation, Grant ECS-0500925, NASA, Grant, NNG04GC11G, SPO #000059778, SPO #00009674, AFOSR, Grant FA9550-06-1-0462. www.rle.mit.edu/media/pr150/34.pdf
47. Beck, M., Hofstetter, D., Aellen, T., Faist, J., Oesterle, U., Ilegems, M., Gini, E., Melchior, H.: Continuous wave operation of a mid-infrared semiconductor laser at room temperature. Science **295**, 301–305 (2002)
48. Evans, A., Darvish, S.R., Slivken, S., Nguyen, J., Bai, Y., Razeghia, M.: Buried heterostructure quantum cascade lasers with high continuous-wave wall plug efficiency. Appl. Phys. Lett. **91**, 071101-1–071101-3 (2007)
49. Bai, Y., Darvish, S.R., Slivken, S., Zhang, W., Evans, A., Nguyen, J., Razeghi, M.: Room temperature continuous wave operation of quantum cascade lasers with watt-level optical power. Appl. Phys. Lett. **92**, 101105-1–101105-3 (2008)
50. Mukherjee, A., Prasanna, M., Lane, M., Go, R., Dunayevskiy, I., Tsekoun, A., Patel, C.K.N.: Optically multiplexed multi-gas detection using quantum cascade laser photoacoustic spectroscopy. Appl. Opt. **47**, 4884–4887 (2008)
51. Van Neste, C.W., Senesac, L.R., Thundat, T.: Standoff photoacoustic spectroscopy. Appl. Phys. Lett. **92**, 234102-1–234102-3 (2008)
52. Taslakov, M., Simeonov, V., Van den Bergh, H.: Line-of-sight data transmission system based on Mid IR quantum cascade laser. Proc. SPIE **6877**, 68770F.1–68770F.10 (2008)
53. Bai, Y., Slivken, S., Darvish, S.R., Razeghi, M.: Room temperature continuous wave operation of quantum cascade lasers with 12.5% wall plug efficiency. Appl. Phys. Lett. **93**, 021103-1–021103-3 (2008)
54. Razeghi, M., Slivken, S., Bai, Y., Gokden, B., Darvish, S.R.: High power quantum cascade lasers. New J. Phys. **11**, 125017(1–13) (2009)
55. Bismuto, A., Gresch, T., Bachle, A., Faist, J.: Large cavity quantum cascade lasers with InP interstacks. Appl. Phys. Lett. **93**, 231104-1–231104-3 (2008)
56. Slivken, S., Evans, A., Zhang, W., Razeghi, M.: High-power, continuous-operation intersubband laser for wavelengths greater than 10 μm. Appl. Phys. Lett. **90**, 151115-1–151115-3 (2007)
57. Vurgaftman, I., Meyer, J.R.: Photonic-crystal distributed-feedback quantum cascade lasers. IEEE J. Quantum Electron. **38**, 592–602 (2002)
58. Bai, Y., Darvish, S.R., Slivken, S., Sung, P., Nguyen, J., Evans, A., Zhang, W., Razeghi, M.: Electrically pumped photonic crystal distributed feedback quantum cascade lasers. Appl. Phys. Lett. **91**, 141123-1–141123-3 (2007)
59. Tredicucci, A., Gmachl, C., Capasso, F., Sivco, D., Hutchinson, A., Cho, A.: A multiwavelength semiconductor laser. Nature **396**, 350–353 (1998)

60. Gmachl, C., Tredicucci, A., Sivco, D., Hutchinson, A., Capasso, F., Cho, A.: Bidirectional semiconductor laser. Science **286**, 749–752 (1999)
61. Owschimikow, N., Gmachl, C., Belyanin, A., Kocharovsky, V., Sivco, D., Colombelli, R., Capasso, F., Cho, A.: Resonant second-order nonlinear optical processes in quantum cascade lasers. Phys. Rev. Lett. **90**, 043902-1–043902-4 (2003)
62. Bengloan, J., Rossi, A., Ortiz, V., Marcadet, X., Calligaro, M., Maurin, I., Sirtori, C.: Intracavity sum-frequency generation in GaAs quantum cascade lasers. Appl. Phys. Lett. **84**, 2019–2021 (2004)
63. Chamberlin, D., Robrish, P., Trutna, W., Scalari, G., Giovannini, M., Ajili, L., Faist, J., Beere, H., Ritchie, D.: Amplification of terahertz radiation in delta-doped germanium thin films in terahertz and gigahertz electronics and photonics IV. Proc. SPIE **5727**, 44–53 (2005)
64. Straub, A., Gmachl, C., Sivco, D.L., Sergent, A.M., Capasso, F., Cho, A.Y.: Simultaneously at two wavelengths (5.0 and 7.5 µm) single mode and tunable quantum cascade distributed feedback lasers. Electron. Lett. **38**, 565–567 (2002)
65. Straub, A., Gmachl, C., Mosely, T.S., Colombelli, R., Troccoli, M., Sivco, D.L., et al.: Tow-wavelength quantum cascade lasers with heterogeneous cascade. In: Proceeding of IEEE Conference on Optoelectronic and Microelectronic Materials and Devices, pp. 141–144 (2002)
66. Gmachl, C., Shu, G., Howard, S.S., Toor, F., Dirisu, A., Malis, O., et al.: Multi-wavelength and nonlinear quantum cascade lasers. In: Proceeding of APS March Meeting (2006). http://meetings.aps.org/Meeting/MAR06/Event/41920
67. Rosencher, E., Fiore, A., Vinter, B., Berger, V., Bois, Ph., Nagle, J.: Quantum engineering of optical nonlinearities. Science **271**, 168–173 (1996)
68. Bai, J., Citrin, D.S.: Optical and transport characteristics of quantum-cascade lasers with optimized second-harmonic generation. IEEE J. Quantum Electron. **43**, 391–398 (2007)
69. Austerer, M., Schartner, S., Pflügl, C., Andrews, A.M., Roch, T., Schrenk, W., Strasser, G.: Second-harmonic generation in GaAs-based quantum-cascade lasers. Physica E **35**, 234–240 (2006)
70. Mann, Ch., Yang, Q.K., Fuchs, F., Bronner, W., Kiefer, R., Köhler, K., Schneider, H., Kormann, R., Fischer, H., Gensty, T., Elsässer, W.: in Quantum Cascade Lasers for the Mid-infrared Spectral Range. Devices and Applications (Advance Solid State Physics 43), ed. by Kramer, B. (Springer, Berlin), pp. 351–368 (2003)
71. Schneider, H., Liu, H.C.: Quantum Well Infrared Photodetectors: Physics and Applications. Springer Series in Optical Sciences, vol. 126. Springer-Verlag, Berlin (2007)
72. Smet, J.H., Fonstad, C.G., Hu, Q.: Intrawell and interwell intersubband transitions in multiple quantum wells for far-infrared sources. J. Appl. Phys. **79**, 9305–9320 (1996)
73. Paiella, R.: Intersubband transitions in quantum structures. McGraw-Hill, New York (2006)
74. Köhler, R., Tredicucci, A., Beltram, F., Beere, H.E., Linfield, E.H., Davies, A.G., et al.: THz semiconductor-heterostructure laser. Nat. (Lond.) **417**, 156–159 (2002)
75. Williams, B.S., Kumar, S., Qin, Q., Hu, Q., Reno, J.L.: Terahertz quantum cascade lasers with double-resonant-phonon depopulation. Appl. Phys. Lett. **88**, 261101-1–261101-3 (2006)
76. Walther, C., Scalari, G., Faist, J., Bree, H., Ritchie, D.: Low frequency terahertz quantum cascade laser operating from 1.6 to 1.8 THz. Appl. Phys. Lett. **89**, 231121-1–231121-3 (2006)
77. Luo, H., Laframboise, S.R., Wasilewski, Z.R., Aers, G.C., Liu, H.C., Cao, J.C.: Terahertz quantum-cascade lasers based on a three-well active module. Appl. Phys. Lett. **90**, 041112-1–041112-3 (2007)
78. Bellotti, E., Driscoll, K., Moustakas, T.D., Paiella, R.: Monte Carlo study of GaN versus GaAs terahertz quantum cascade structures. Appl. Phys. Lett. **92**, 101112-1–101112-3 (2008)
79. Harima, H.: Properties of GaN and related compounds studied by means of Raman scattering. J. Phys. Condens. Matter. **14**, R967–R993 (2002)
80. Jovanovic, V.D., Indjin, D., Ikonic, Z., Harrison, P.: Simulation and design of GaN/AlGaN far-infrared ($\lambda \sim 34$ µm) quantum-cascade laser. Appl. Phys. Lett. **84**, 2995–2997 (2004)
81. Sun, G., Soref, R.A., Khurgin, J.B.: Active region design of terahertz GaN/Al$_{0.15}$Ga$_{0.85}$N quantum cascade laser. Superlattices Microstruct. **37**, 107–113 (2005)

82. Harrison, P., Indjin, D., Jovanovic, V.D., Mircetic, A., Ikonic, Z., Kelsall, R.W., et al.: A physical model of quantum cascade lasers: application to GaAs, GaN and SiGe devices. Phys. Status Solidi A **202**, 980–986 (2005)
83. Scalari, G., Sirigu, L., Terazzi, R., Walther, C., Amanti, M.I., Giovannini, M., et al.: Multiwavelength operation and vertical emission in THz quantum-cascade lasers. J. Appl. Phys. **101**, 081726-1–081726-5 (2007)
84. Indijan, D., Ikonić, Z., Jovonović, V.D., Harrison, P., Kelsall, R.W.: Relationship between carrier dynamics and temperature in terahertz quantum cascade structures: simulation of GaAs/AlGaAs, SiGe/Si and GaN/AlGaN devices. Semicond. Sci. Technol. **20**, S237–S245 (2005)
85. Ferreira, R., Bastard, G.: Evaluation of some scattering times for electrons in unbiased and biased single- and multiple-quantum-well structures. Phys. Rev. B **40**, 1074–1086 (1989)
86. Abbar, B., Bouhafs, B., Aourag, H., Nouet, G., Ruterana, P.: First-principles calculations of optical properties of AlN, GaN, and InN compounds under hydrostatic pressure. Phys. Status Solidi B **228**(2), 457–460 (2001)
87. Capasso, F., Tredicucci, A., Gmachl, C., Sivco, D.L., Hutchinson, A.L., Cho, A.Y., et al.: High-performance superlattice quantum cascade lasers. IEEE J. Sel. Top. Quantum Electron. **5**, 792–807 (1999)
88. Ajili, L.: Quantum cascade lasers at terahertz frequencies. These, Docteur En Sciences, Institut de physique Universite de Neuchatel (2007)
89. Colombelli, R., Straub, A., Capasso, F., Gmachl, C., Blakey, M.I., Sergent, A.M., Chu, S.N., West, K.W., Pfeiffer, L.N.: Terahertz electroluminescence from superlattice quantum cascade structures. J. Appl. Phys. **91**(6), 3526–3529 (2002)
90. Rostami, A., Baghban, H., Rasooli Saghai, H., Noori, M.: Linear frequency-doubling in dual Mid-IR-wavelength quantum cascade laser active region. Superlattices Microstruct **45**, 134–142 (2009)
91. Rostami, A., Mirzaei, B., Baghban, H.: Two-wavelength THz quantum cascade laser with highly enhanced temperature characteristics. In: Proceedings of SPIE-OSA-IEEE Asia Communications and Photonics, SPIE-OSA-IEEE, ACP **7631**, 76310N-76310N-8 (2009)

Index

1/f noise, 93

A
Absorption coefficient, 73, 74, 92, 100, 101, 113, 116, 134, 135, 140, 141, 144, 145, 147, 159, 183, 211
Absorption spectra, 6, 71, 108
Acceptors, 66
Acoustic bolometer, 19
Acoustical phonons, 95
Active region, 30, 43, 92, 106, 111, 119, 120, 122, 123, 125, 126, 180, 182, 183, 192, 197, 200–203, 205–208, 211, 220–224, 226, 227, 229, 232
Airy disk, 16
All-optical techniques, 43, 70
Ambient background thermal noise, 16
Amorphous, 70
Amorphous polycarbonate (APC), 71
Analog-to-digital (ADC), 31
Anderson localization, 13
Antenna, 7, 9, 18, 25, 27, 39, 41, 42, 46, 52, 59, 62–67, 70, 123
Aperture, 3, 32, 35, 36
Aqueous environment, 15
Artificial atoms, 123
Artificial magnetic, 53
Artificial materials, 53
Artificial molecules, 123
Astrophysics, 1
Atom-field interaction Hamiltonian, 153
Atomic force microscope image, 121
Atomic vapour systems, 153
Autler–Townes doublet, 163, 164, 166, 171, 172, 175, 176, 183
Autocorrelation, 68

B
Background blackbody radiation, 94
Background limited IR performance (BLIP), 95
Backward wave oscillators (BWO), 49
Ballistic mode, 24
Band offset, 135, 142, 148, 179, 216
Bessel Functions, 140
Bianisotropy, 55
Biological tissues, 11
Biomedical, 1
Biomolecular, 11
Bipolar laser, 191
Birefringence, 10, 44, 73
Blackbody emissivity, 94
Blackbody spectral density, 104
Blocking barrier, 105, 106, 125, 221
Bolometers, 19, 22, 28, 29, 38
Boltzmann constant, 161, 174, 221
Boltzmann transport equation, 213
Bose–Einstein, 97, 221
Bound-to-continuum transition, 202, 203
Bow-tie antenna, 27, 67
Buried ridge geometry, 215

C
Calorimeter, 19
Cancer, 11
Canonical momentum operator, 153
Capping layer, 132–134
Capture probability, 92, 103–106, 142
Carbon nanotube (CNT) quantum dot (QD), 32
Carcinotrons, 40, 42
Carrier lifetime, 52, 63, 64, 221
Carrier momentum relaxation time, 64
Carrier relaxation mechanism, 192

C (*cont.*)
Cascaded multipliers, 48
Cascaded nonlinear processes, 57
Cavity-enhanced, 60
CCD camera, 10
Centered defect quantum dot, 134, 135
Chemical reactions, 14
Chemical vapor deposition, 121
Chirped superlattice, 201, 202
Chromophore, 71
Cladding layer, 204–206, 208, 211, 212
Coherent detection, 44, 68, 123
Coherent terahertz imaging, 44
Coherent trapping, 155, 156, 160
Collection efficiency, 101
Collective modes, 15, 31
Composite bolometers, 18, 19, 21
Confinement factor, 200, 202, 203, 207, 208, 215, 224, 229
Conjugated semiconducting polymers, 70
Copolymer, 71
Coulomb blockade, 33
Coupled quantum dots, 149, 150, 152
Coupling coefficient, 211
Cross-gain effects, 223
Cryogenic cooling, 16, 17, 40, 42
Crystal polar axis, 73
Current limited, 94

D
Dark current noise, 93–95, 104
Dark states, 155
DAST, 45, 70, 72–77
DCDHF type dye, 71
de Broglie wavelength, 120
Decay rates, 155, 174, 181
Density matrix formalism, 173, 181
Density operator, 154
Dental caries, 12, 13
Dephasing decay rates, 174
Dephasing scattering, 213, 215
Depopulation, 198, 203, 213, 226
Destructive quantum interference, 156
Detectivity, 91, 93–95, 101, 102, 104, 106, 109, 110, 114, 124, 126, 129, 134, 135, 142, 146, 148, 183
Detuning, 161, 162, 165–167, 169, 177–179
Difference frequency generation (DFG), 56, 57, 60, 61, 77
Differential spectroscopy, 219
Diffraction limit, 34, 219
Diffusion-cooled HEBs (DHEB), 37

Dipole approximation, 153
Dipole-type antenna, 63
Direct detectors, 3, 16, 18
Dislocation, 120
Dispersion theory, 207
Distributed Bragg reflector (DBR), 52, 69
DNA, 4, 11
Dome shaped, 149–151
Donors, 66, 130
Dot-well system, 128, 133
Double barrier, 105, 125, 127, 136, 140–142, 145, 147, 148
Double-coupled quantum well (DCQW), 67
Down-converter, 16, 40
Drude model, 207
Dual-wavelength imaging, 107

E
Eigenfunctions, 139
Eigenvalues, 139
Elastic scattering mechanism, 194
Electro-magnetic, 9
Electrochemical potential, 15
Electromagnetically induced transparency (EIT), 92, 152, 179
Electron microscope, 42
Electron spin resonance (ESR) spectroscopy, 15
Electron–electron interaction, 22, 23, 95, 222
Electronic coupling, 123
Electronic split ring resonator (eSRR), 55
Electronic transport, 95, 105
Electron–phonon scattering, 30, 124, 194
Electro-optic crystal, 10
Electro-optic rectification, 45
Electro-optical sampling (EOS), 44
Elongated pyramids, 122
Emission efficiency, 64, 66
Emission of LO phonons, 96
Energy conservation, 74, 222
Energy level splitting, 33, 149, 151, 152
Environment temperature, 152, 161, 165, 168, 170–172, 176
Epitaxial growth, 69, 121, 149
Equation of motion, 154
Escape probability, 92, 95, 103
Esophageal cancer, 11
Evanescent field, 32, 35
Excited state, 82, 103, 107, 111, 113, 142, 144, 145, 148, 196, 197, 231
Excitonic charge oscillations, 67
External cavity laser diode (ECLD), 56
External quantum efficiency, 101

Index

F
Fabry–Perot, 210
Far field, 13, 217, 219
Femtosecond lasers, 14, 42, 118
Femtosecond pulses, 42, 60
Fermi's golden rule, 192, 221
Fermi–Dirac, 97, 98, 100, 141
Ferroelectric materials, 20
Fingerprinting, 9
Focal plane arrays, 105
Folded waveguide travelling wave tube (FWTWT), 51
Fourier transform infrared spectroscopy, 24
Four-wave mixing, 157
Four-zone structure, 106
Fowler–Nordheim tunneling, 118
Free-carrier absorption, 119
Free-electron absorption, 22
Free-electron laser (FEL), 42, 119
Frequency-tunable terahertz detectors, 29
Fröhlich interaction, 194
FWHM, 129, 148, 199, 219, 231

G
Gain coefficient, 199, 220, 222, 223
Generation-recombination noise, 94, 105
Global transition rate, 96, 107, 182, 183
Golay cells, 19
Grating tuned CO pump lasers, 41
Ground state, 67, 82, 95, 107, 109, 111, 118, 130, 135, 144, 151, 152, 172, 178, 196, 197
Guest–host configuration, 71
Gunn, 24, 40, 42, 50, 84

H
Hamiltonian, 96, 153–155, 158, 159, 180, 193, 194
Hanckel, 140
Harmonic interaction, 193
Heavy-hole band, 42, 150
Heisenberg uncertainty principle, 17
Helium-cooled, 19, 68
Hertzian dipole-type antenna, 63
Heterodyne detectors, 3, 16
Heterodyne receiver, 3
Heterodyne semiconductor, 16
Heterodyne superconductor, 17
Heterojunction, 28, 123
Heterostructure, 25, 30, 33, 35, 43, 111, 126, 127, 129, 136, 179, 192

Heterostructure bipolar transistors (HBTs), 25
High electron mobility transistors (HEMTs), 25
High power quantum cascade lasers, 215
High-Tc bridge, 65
Hot electron bolometer, 17, 23, 36

I
Idler, 49, 58, 60, 74
IMPATT, 24, 40, 42
In vitro, 11
In vivo, 1, 11, 12
Incoherent detectors, 123
Index coupled QC lasers, 211
Infrared countermeasures, 215
Injection efficiency, 198, 200, 203, 224, 228
Injector, 197, 198, 200, 203, 213, 216, 220, 221, 223, 226
Inorganic, 15, 45, 70, 73
Interaction Hamiltonian, 153, 193
Interband transitions, 193
Intercascade, 97, 113
Interdigitated electrodes, 52
Interfaces roughness, 174
Interferential state, 181
Interlaced, 202, 203
Inter-miniband transitions, 43, 200
Interstellar, 2, 3
Intersubband relaxation time, 101, 124
Intersubband scattering mechanism, 195
Intersubband transitions, 41, 43, 91, 100, 106, 115, 135, 192, 193, 219
Intracascade, 97
Intracavity frequency mixing, 219
Ion implantation, 136, 137
Ionized-impurity scattering, 194
IR-pumped gas lasers, 41

J
Johnson noise, 93, 95, 100–102, 104, 114, 182
Johnson noise limited detectivity, 101, 102, 182
Jones, 93
Josephson-effect, 17

K
Klystrons, 40, 49, 51
KTP crystals, 74

L

Lattice-mismatch, 121
Leakage current, 228
Lens shaped, 122
Light-hole band, 42
$LiNbO_3$, 41, 60, 65
Linewidth enhancement factor, 192
LiTaOs, 76
Log-periodic antenna (LPA), 39
Long-integration-time radiometric
    techniques, 16
Longitudinal optical (LO) phonons, 67, 128, 192
Lorentz–Drude theory, 207
Lorentz-gauge vector potential, 193
Lorentzian broadening, 141
Loss, 13, 16, 40, 49, 60, 61, 75, 198–200, 203, 205–209, 211, 215, 217, 224, 229, 231
Loss-coupled QC-DFB lasers, 211

M

Macromolecules, 14
Magnetic resonance imaging, 11
Mammography, 11
Martin–Puplett (MP), 68
Material gain, 214, 231
Matrix element, 96, 97, 101, 141, 143, 157, 181, 193, 194, 199–201, 219, 221, 222, 224, 231
MBANP, 70, 77
Medical diagnostics, 6, 91
Medical imaging, 9, 91
Metal–insulator–metal (MIM) diode, 25
Metal–organic chemical
    vapor deposition, 121
Metal–Oxide–Metal (MOM), 27
Metal–semiconductor–metal
    structure, 208
Metamaterials, 53, 55
Michelson interferometer, 1
Microbolometers, 19
Microcavities, 69
Microstrip, 47, 209
Microwaves, 9
Miniband, 43, 105, 130, 200, 201, 203, 220, 227
Minigaps, 196, 201
Minimal-coupling Hamiltonian, 153
Modal effective refractive index, 229
Modal gain, 200
Mode-locked laser, 46
Molecular beam epitaxy, 121, 125, 126, 192

Molecular gas lasers, 42
Momentum conservation, 74, 222
Monolithic microwave integrated circuit
    (MMIC), 40
Monte Carlo simulation, 107, 213
Multiband THz detection, 31
Multicolor operation, 219, 226
Multi-level atomic systems, 152
Multi-octave antenna, 52
Multiple quantum-well
    (MQW), 192, 212
Multi-wavelength detection, 107

N

Nanodevices, 29
Nanoelectronic technologies, 32
Near-field imaging, 34
Near-field terahertz detector, 32
Near-infrared, 13, 22, 43–46, 60–62
Negative refractive index, 53
Noise equivalent power, NEP, 93
Noise gain, 94, 102
Noise-equivalent temperature
    difference, 104, 148
Non-collinear phase-matched
    mixing, 56
Nonequilibrium Green's function
    approach, 214, 215
Non-ionizing, 1, 5
Nonlinear materials, 61, 65
Nonlinear spectroscopy, 15
Nonlinear susceptibilities, 157
Non-radiative transitions, 192, 194, 197, 228
Normal incidence light, 30

O

Off-axis parabolic mirror, 44, 58
Offset-frequency-locked CW lasers, 41
One-dimensional confinement, 120
Optical communications, 91
Optical gain, 103, 104, 134, 145, 196, 198, 226, 231
Optical parametric oscillation
    (OPO), 41, 56, 57
Optical polarization, 181
Optical stark effect, 172
Optically-pumped far-IR lasers, 6
Organic materials, 1, 65, 70
Organic photoconductive dipole antennas
    (OPDAs), 73
Oscillator strength, 194, 201, 203, 212
Overall quantum efficiency, 92

Index

P

Parametric process, 42, 58, 59
Peak gain, 231
Permeability, 13, 53, 101
Phase-matching condition, 57, 62, 66, 73
Phonon bottleneck, 135
Phonon modes, 15
Phonon scattering process, 174, 196, 203, 220, 222
Phonon-cooled HEBs, 37
Photo-acoustic spectroscopy, 215
Photo-carrier lifetime, 63
Photocell, 24
Photochemical electron transfer, 15
Photoconductive, 9, 12, 21, 27, 42, 51, 52, 59, 62, 70, 92, 94, 101, 105, 114, 119, 123
Photoconductive dipole antennas, 42
Photoconductor, 9, 29, 40, 41, 71, 94, 104
Photocurrent, 9, 41, 63, 66, 117–120, 125, 172
Photo-Dember effect, 46
Photodiode, 16, 44, 68
Photoemission detectors, 123
Photon lifetime, 224, 229
Photon noise, 93, 94
Photon-assisted tunneling, 17, 32, 196, 197
Photonic crystal cavities (PhCNC), 60, 61
Photonic crystal distributed feedback (PCDFB), 217
Photoresponse, 119, 124, 125, 130, 131
Photovoltaic, 95, 100, 104–107, 182
Picoseconds laser pulsing, 40
Piezoelectric potentials, 149
Planar diode mixers, 16
Plasma fusion, 4
Plasmon-enhanced dielectric waveguide, 205
Plasmonic resonant, 25
Pockels effect, 44, 68
Polar molecules, 78
Polar optical phonons, 65, 66, 200, 221
Pollution monitoring, 210
Poly (*p*-phenylene vinylene) (PPV), 71
Polyacethylene, 71
Polymers, 14, 61, 62, 70–72
Population inversion, 197–201, 212, 226, 231, 232
Propagating mode, 199
Propagation gain, 231
Pulsed mode operation, 215, 216
Pyramids, 122
Pyroelectric detectors, 20

Q

Q-switched Nd:YAG laser, 41, 74
Quadratic detection, 117, 119
Quantum cascade detector, 92, 95, 104
Quantum cascade lasing, 40
Quantum dot intersublevel photodetector (QDIP), 30
Quantum dot island, 136, 137
Quantum dot photodetector, 30, 123, 125, 129, 134
Quantum dot terahertz photodetector, 92
Quantum dots, 29, 30, 32, 103, 120–124, 126, 127, 129, 130, 132, 134, 145, 147–150, 152
Quantum efficiency, 92, 101, 103, 105, 145, 149, 157
Quantum point contacts (QPCs), 29
Quantum ring infrared photodetectors (QRIPs), 132
Quantum size effect, 134
Quantum tunneling, 27
Quantum well, 43, 67, 91, 92, 105–107, 111, 113, 115, 120, 123, 126, 135, 152, 168, 172, 174–176, 179–181, 192, 195, 201, 221, 226
Quantum well infrared photodetectors (QWIPs), 105, 124
Quantum wires, 120
Quasi-continuum, 107
Quasi-optical coupling method, 39
Quasi-particles, 17, 67

R

Rabi frequencies, 156, 159, 163, 164, 174, 180, 182
Radiative transitions, 191, 192
Radiometric techniques, 16
Random lasers, 13
Random media, 13
Rayleigh scattering, 8
Real-time THz imaging, 34
Recycled, 197
Relaxation matrix, 155
Resonant phonon structure, 203
Resonant tunneling diodes (RTDs), 30, 40, 42
Resonant tunneling quantum-dot infrared photodetector, 125
Responsivity, 17, 21, 30, 91, 92, 95, 101–104, 109, 113–115, 123—125, 127, 129, 133–135, 142, 145–148
Reststrahlen band, 119
Rollover current, 215, 217
Room-temperature, 16, 17, 43, 53, 226

R (*cont.*)
Rotating-wave approximation, 155, 159, 173, 180
RT-SCDQD, 134–136, 141, 145–148, 189
Rydberg transitions, 12

S
Satellite peaks, 33
Scattering lifetimes, 228
Schottky barrier diode, 16, 39
Schottky diode mixer, 16, 123
Schrödinger equation, 136, 139, 151, 174
Schrödinger–Poisson analyze, 224
Second-order susceptibility, 41, 115
Self-assembled quantum dots, 121, 122
SEM image, 25, 27, 28, 217
Semi-classical Boltzmann approach, 215
Semiconductor lasers, 42, 56, 192
Semiconductors, 9, 28, 43, 45, 46, 60, 65, 66, 69, 70, 91, 95, 121, 123, 150, 192, 199
Semi-insulating, 132, 209, 210
Semimetals, 67
Sensor networking, 91
Sequential tunneling, 135, 198
Shot noise, 93, 104
Signal-to-noise-ratio, 11, 93
Silicon waveguide, 60, 61
Single electron transistor (SET), 19, 33
Single quantum well (SQW), 67, 106
Single-cycle electric pulse emitter, 68
SIS mixer, 17
Slope efficiency, 200, 201, 211, 215, 233
Smith–Purcell emitter, 42
Spacer, 111, 122, 141, 145, 148–151
Spherical harmonics, 137
Split ring resonator (SRR), 53–55
S-polarized, 128
Static relative permittivities, 96
Stepped quantum well, 113
Strain, 121
Strain–relaxation, 121–123, 125, 136, 149–151, 215
Stranski–Krastanow, 121, 122
Sub-millimeter, 1
Superconductor–insulator–superconductor (SIS), 17
Superlattice, 106, 107, 130, 132, 196, 201, 202, 215
Surface plasmons, 27, 201
Surface-plasmon waveguide, 206, 208
Surge current, 65, 66, 69

Susceptibility, 41, 45, 60, 140, 158, 160, 161–168, 170, 174, 181, 219
Synchrotron sources, 42

T
Terahertz detectors, 3, 16, 17, 29, 32, 127, 129, 170
Terahertz imaging, 9, 32, 44
Terahertz radiation, 1–3, 5, 7–9, 17, 29, 42, 43, 45, 51, 52, 56–59, 61, 62, 127, 149, 152, 181
Terahertz spectroscopy, 4, 5
Terahertz tube sources, 41
Terahertz-waveguide, 208
Thermal backfilling, 201, 226
Thermal background dark current, 30
Thermal broadening, 135, 152
Thermal equilibrium, 23, 69, 97
Thermal imaging, 91
Thermal noise limited detector, 16
Thermionic emission, 103, 124, 135, 148, 152, 221
Thermistor, 22
Threshold current, 199, 211, 215, 217, 218, 225, 231
THz camera, 34
THz dielectric spectroscopy, 24
THz microscopy, 34
Time domain spectroscopy, 44
Time-dependent density-matrix equations, 173, 180
Transfer matrix method, 139
Transmission coefficient, 74, 103, 104, 139, 140, 159, 161, 162, 164–166, 168, 171, 174–177, 181, 182
Transmission electron microscopy (TEM), 122
Transport ladder, 111, 113, 179
Traveling wave terahertz detector, 25, 27
Travelling Wave Tubes (TWT), 49
Tunable plasma wave -HEMT THz detector, 27
Tunnel junction mixer, 17
Tunneling probability, 67, 145
Two-dimensional electron gas (2DEG), 25, 35, 121
Two-photon absorption, 92, 115

U
Ultrasonography, 11
Ultraviolet, 28
Ultra-wide tenability, 73
Up-converters, 40

V

Vacuum tubes, 50
Vertical coupling, 123
Vibrational and rotational, 8

W

Wall plug efficiency (WPE), 215
Wave function, 97, 98, 111–113, 123, 137, 140, 141, 143, 150, 154, 156, 172, 179, 226
Wavelength division multiplexing, 219
Wet chemical etching, 211, 212
Wetting layer, 121, 143, 149
WKB approximation, 221
Wollaston polarization (WP), 44

X

X-ray, 8, 11, 12

Z

Zero-dimensional confinement, 120